Better City, Better Life

Better City, Better Life

Lin Chau Ming
Wenhua Wang
Renata Cardoso Magagnin
(Orgs.)

CULTURA ACADÊMICA
Editora

Cultura Acadêmica
Praça da Sé, 108
01001-900 – São Paulo – SP
Tel.: (0xx11) 3242-7171
Fax: (0xx11) 3242-7172
www.culturaacademica.com.br
www.livrariaunesp.com.br
feu@editora.unesp.br

CIP – BRASIL. Catalogação na publicação
Sindicato Nacional dos Editores de Livros, RJ

B466

Better city, better life / Organização Renata Cardoso Magagnin, Lin Chau Ming, Wenhua Wang. – 1.ed. – São Paulo: Cultura Acadêmica, 2016.

ISBN: 978-85-7983-828-6

Texto em inglês

1. Meio ambiente. 2. Preservação ambiental. 3. Qualidade de vida. I. Magagnin, Renata Cardoso. II. Ming, Lin Chau. III. Wang, Wenhua.

16-38593 CDD: 577
CDU: 502.1

Editora afiliada:

Asociación de Editoriales Universitarias de América Latina y el Caribe

Associação Brasileira de Editoras Universitárias

Sumário

Introduction and Overview 9

1. Urban Housing and Environment: Challenges for Sustainability 13
   Laura Machado de Mello Bueno

2. Ecolinguistics and Environmental Education in Brazil......... 25
   José Marcelo Freitas de Luna

3. Water Resources in China: situation, measures and actions . 35
   Jiahui Shao

4. Urban planning in Brazil 49
   Ítalo Stephan

5. Sustainability, Development and Well-being at Work........... 63
   Inês Monteiro

6. Strengthening Regional Capacities for Providing Remote Sensing Decision Support in Brazil in the Framework of Top China 71
   Humberto Alves Barbosa

7. One mode of Urban Agriculture: Organic Vegetable Production and Marketing 83
Huang Danfeng e Wang Xiaoli

8. Plant Biotechnology: An Approach for Pesticide Reduction in Agriculture 97
Eliane Romanato Santarém

9. Environmental Preservation and Protection of the Built Environment in Megacities 107
Clovis Chiezzi Seriacopi Ferreira

10. How Information Technology Can Improve Quality of Life for Elderly People 127
Anita Maria da Rocha Fernandes

11. Poverty in metropolitan areas and conditional cash transfer programs in Brazil: the Campinas (SP) experience 139
André Pires

12. An Indispensable Architectural Experience: João Filgueiras Lima – Lelé 153
Anália Maria Marinho de Carvalho Amorim

13. Sustainable Energy Sources in the Global Warming Scenario 169
Adir Janete Godoy dos Santos

14. Better use of scientific knowledge for green, clean and sustainable cities 181
Luciana Pinto Sartori

15. Bacterial regrowth and its control in drinking water distribution system 197
Yiming Liao e Xiaohui Bai

16. Measurement of Mercury in Flue Gas Based on an Aluminum Matrix Sorbent and Mercury Removal by Existing Pollutant Control Devices of Six Coal-fired Power Plants in China 209
Wang Wenhua

17. Integrated watershed management for sustainable land use in Brazil: Soil and Water Resources 229
Teresa Cristina Tarlé Pissarra

18. Better City, Better Life – Minimum Sustainable Architecture Minimum Sustainable Architecture........ 245
Silvio Stefanini Sant'Anna

19. The Sewage Treatment Plant with a Case Study in Cachoeiro de Itapemirim, ES, Brazil 257
Sandra Novaes Coelho

20. Using Role Play to Build a Common Sustainable Development Agenda for Brazil-China: the Case of Top China Program........ 271
Rogério César Pereira de Araújo e Leda Nakashiro Makiya

21. Three-Dimensional Air Quality Models........ 289
Ping Liu

22. Urban and Regional "Green" Mobility: Focus on Transportation 299
Nadja Lisboa da Silveira Guedes

23. Conservation of Mangroves in Urban Areas........ 311
Manuel Antonio de Andrade Furtado-Neto

24. Risk, Modernity and Networks: The Risk of Modernity and the Emergence of Networks in Industry........ 323
Luciano Modes

25. Urban conservation in Brazil: an overview........................... 333
Leonardo Barci Castriota

26. Risk and Climate Change in Brazilian Coastal Cities (The case of São Paulo State)... 345
Leila da Costa Ferreira

Introduction and Overview

The book *Better City, Better Life* brings together papers from different disciplines of researchers who have the theme Sustainability in common. Another common characteristic refers to this theme's participation in the Top China International Mobility Program.

This book is intended to reflect on current planning strategies and growth of cities from the perspective of sustainable development. These reflections approach the spatial, economic, political, social, cultural and environmental model. This book is divided into twenty-seven chapters and subdivided into the following themes: *Sustainable Cities*, *Environmental Sustainability* and *Social and economic aspects of sustainability*.

The theme *Sustainable Cities* includes papers related to sustainability in architecture and cities based on the following approaches: sustainability of buildings, housing, urban planning, transportation planning, legislation for city planning, management of cities, among other related topics.

The chapter, "Better City, Better Life: Minimum sustainable architecture", aims to understand the concept of sustainability and its application in architecture. "Urban housing and environment: challenges for sustainability" discusses the process of urbanization and housing conditions in Brazil through the global dimension and the social and environmental crisis. The chapter, "An indispensable architectural experience: João Filgueiras Lima – Lelé", aims to present an architectural design experience that unites the design of architectural and urban space with the constructive logic of cities. "Urban conservation in Brazil: an overview" shows the history of urban conservation in

Brazil, focusing on its early days, while trying to understand what in particular led intellectuals and politicians of the time to propose the preservation of complete sets in the 1930s. The chapter, "Urban planning in Brazil", aims to reflect on the process of urbanization in Brazil. "Better use of scientific knowledge for green, clean and sustainable cities" intends to reflect on the importance of cities have in planned public spaces that are sustainable and generate quality of life for the population. The chapter, "Environmental preservation and protection of the built environment in megacities", reflects on the challenges that cities face between urban development and the preservation of the environment. The chapter, "Green spaces in metropolitan area: spatial distribution and ecosystem service quantification", analyzes urban vegetation in the metropolitan area. The case study in Pudong District, Shanghai, aims to show the nature of spatial distribution and ecosystems in urban green spaces. Finally, the chapter "Urban and regional" green "mobility: focus on transportation" aims to reflect on the theme sustainable transport for cities in terms of the existence of good Urban Mobility.

The second theme, "Environmental sustainability", refers to sustainability issues related to the environment. The papers discuss the following topics: energy, climate, natural resources, infrastructure networks and smarter farming practices that are not harmful to health or the environment, among other related topics.

The chapter, "Energetic sources sustainability in the global warming scenario", analyzes the perspectives of the energy matrix in Brazil through the following topics: demand, environment, regulation, technology and strategy. "Risk and climate change in Brazilian coastal cities (The case of São Paulo State)" investigates some of the political and institutional structures that have an effect on climate change in the coastal cities of São Paulo. The chapter, "Three-Dimensional Air Quality Models", refers to the study of a three-dimensional model to analyze the transformation and transport of air pollutants in the atmosphere. "Conservation of Mangroves in urban areas" shows a discussion on the conservation of mangroves. The paper, "Water resources in China – situation, measures and actions", analyzes the problems related to water resources in China. "Bacterial regrowth and its control in drinking water distribution system" analyzes bacterial growth and control in the drinking water distribution system in

China. The chapter, "The sewage treatment plant with a case study in Itapemirim, ES, Brazil", shows a Brazilian example regarding the treatment of municipal sewage. "Integrated watershed management for sustainable land use in Brazil: soil and water resources" shows a discussion on the management of land use to improve urban, industrial and rural activities. The Chapter, "Strengthening Regional Capacities for providing remote sensing decision support in Brazil in the framework of Top China", aims to present basic information about remote sensing tools. "Measurement of mercury in flue gas based on an aluminum matrix sorbent and mercury removals by existing pollutants control devices of six coal-fired power plants in China" shows the mercury pollution-related problems at coal plants in China. The chapter, "One mode of urban agriculture: organic vegetable production and marketing", aims to analyze the characteristics of organic plant development and the challenges and opportunities for organic agriculture. And the chapter, "Plant Biotechnology: an approach is reduction pesticides in agriculture", shows a study on reducing pesticides in agriculture.

The third part denominated "*Social and economic aspects of sustainability*" brings more specific issues related to sustainability. These papers are concerned with education, work, income, technology and industry. The chapter, "Ecolinguistics and environmental education in Brazil", shows a didactic experience regarding the approach towards the sustainability issue by a group undergoing heterogeneous training, which involved Brazilian and Chinese students. The chapter, "Using role plays to build a common sustainable development schedule for Brazil-China: the case of Top China Program", aims to identify and discuss the issues surrounding sustainable development in Brazil and China. The chapter, "Sustainability, development and well-being at work", analyzes the question of sustainability in the process of changing modes of production enterprises and their impact on people's lives and the planet. The chapter, "Poverty in metropolitan areas and conditional cash transfer programs in Brazil: the Campinas (SP) experience", aims to analyze the income transfer program instituted by the Brazilian federal government. The chapter, "How information technology can Improve elderly people quality of life", shows some examples about how information technology can improve the quality of life of older people in major cities such as Shanghai and Sao Paulo. And the last chapter, "Risk, modernity and networks: the risk in modernity and the emergence of networks in industry", reflects on the risk that exists in modern times with the emergence of networks in industry.

The themes presented in this book are diverse, however, based on the diversity issue, the reader can develop a more integrated point of view regarding research in the area of sustainability.

The book *Better City, Better Life* is designed for researchers, graduate students and professionals in fields related to Architecture and Urban Planning, Urban and Regional Planning, Engineering, Biology, Ecology, and other related fields. It is expected that the presented research can contribute towards the training of these professionals.

1

Urban Housing and Environment: Challenges for Sustainability

Laura Machado de Mello Bueno[1]

Introduction

This paper is based on a lecture given during the Top China program in 2010, 20th July, at Shanghai Jiao Tong University. Firstly, some highlights about Brazil in the world; its population, urbanization process and housing conditions as well as its global dimension and social and environmental crisis are introduced. Secondly, the water and the housing system, as a key factor to welfare and the real possibilities of a sustainable city, are discussed. And finally, some guidelines and challenges related to knowledge and the production of urban space are presented.

1 Professor at the Pontifical Catholic University of Campinas; Centre for Environmental Studies and Technology. Architect and planner, professor at the Faculty of Architecture and Planning and at the Graduate Program in Urban Planning at PUC Campinas. Leader of the Research Group of Water in the urban environment. Currently coordinates a collaborative research project on Climate Change and Urban Form in the metropolitan regions of Rio de Janeiro and Campinas.

Some highlights about Brazil in the world

Brazil is located in South America. The country's population is about 190 million people (2010 census) and its territory covers about 8,500,000 km^2. The country has a vast shoreline along the Atlantic Ocean. The country borders Venezuela, Guyana, Suriname, French Guiana, Colombia, Peru, Bolivia, Paraguay, Argentina and Uruguay (Figure 1). We are the only country in Latin America that speaks Portuguese and that uses this one language all over the entire country.

Figure 1. Brazil and the Distribution of Urban Population

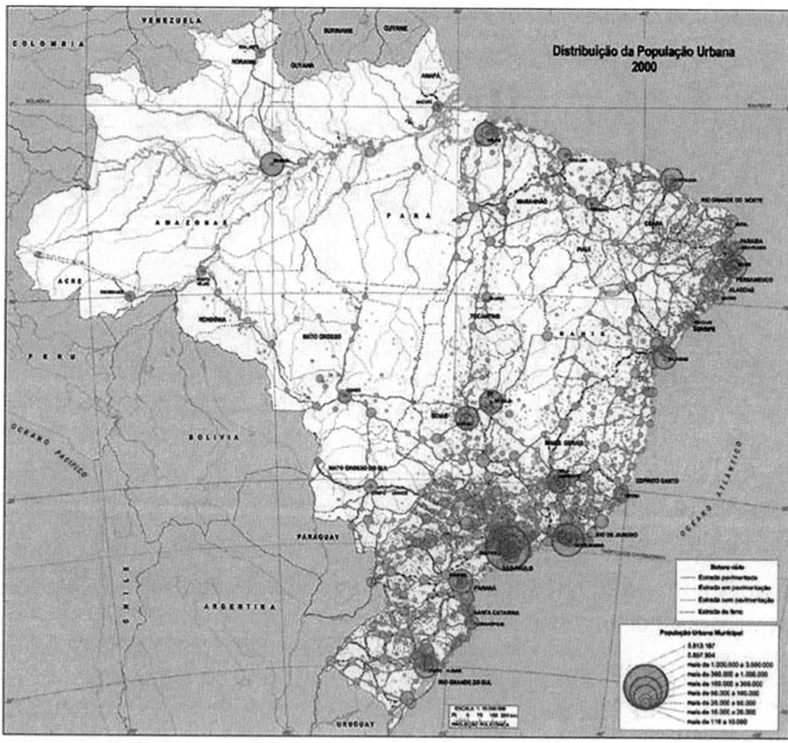

Source: IBGE 2010.

The Brazilian population has grown and changed a lot over the last 60 years, specifically, it has multiplied four times. Nowadays, 160 million inhabitants (85% of the total population) live in cities (IBGE, 2010).

Due to this rapid urban growth, the lack of planning and the fragile administrative structures in the municipalities, cities in Brazil do not have proper infrastructure, sanitation or other social facilities.

Besides this problem, the distribution of the population in Brazil's territory is very unequal. Most of the population and the largest cities and metropolises exist on the coast. This situation was a result of Brazil's history of colonization by the Portuguese Kingdom, which used our territory to develop an exportation system of several raw materials.

The territorial organization of Brazil is based on more than 5000 municipalities, organized accross 27 states. All are federative units that make up the Federative Republic of Brazil. In 1988, a new National Constitution was approved, following 24 years of a military dictatorial regime. The municipalities are responsible for the land use and control, urban policies, as well as the local services, such as sanitation and traffic.

Most part of the 5565 municipalities have small populations. There are 2515 municipalities with less than ten thousand inhabitants. On the other hand, there are 280 municipalities that are currently home to 80 million inhabitants.

The metropolitan areas were formed throughout the country by means of the migration process. Today there are 27 metropolitan areas, which are mainly close to the Atlantic Ocean and important ports – São Paulo, Rio de Janeiro, Salvador, Recife. Only three metropolises are in the countryside – Brasília, Manaus and Belém (Figure 1).

China and Brazil have a similarity – the size of the country (China – 9,536,499 km^2; Brazil – 8,515,492 km^2). However, the size of the two countries' populations is very different (China –1.35 billion people, Brazil – 190 million). Brazil has 84% of its population residing in cities and villages, while China has 45%. The number of cities with more than 750,000 inhabitants in China is much more significant – 143 cities, while in Brazil there are only 24 large cities with more than this number.[2]

Other important differences are the social and economic structures, which are related to the historical insertion on the world political map. This point can be seen by analyzing the Figure 2 above, which shows

---

2 Fonte: UN–HABITAT State of The World's Cities-2010/2011, 2010 e *http://www.census.gov/ipc/www/* consulta em 18 de maio e 6 de junho de 2011.

the share of income or expenditure of selected countries. Brazil has a huge income concentration that makes it one of the most unequal countries in the world.

Figure 2. Share of income or expenditure %

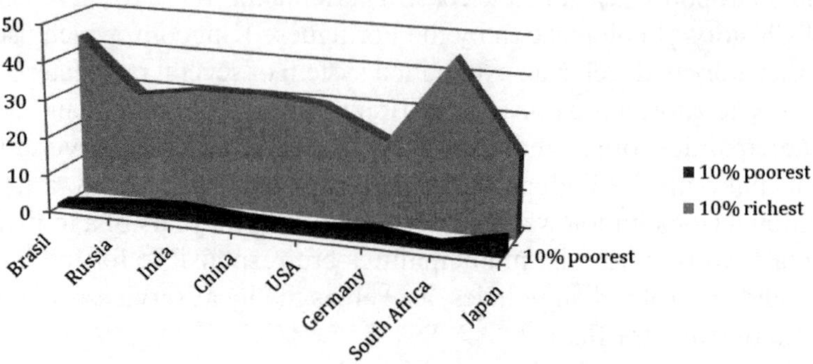

Source: Human Development Report UNPD-2009.

Urban Housing: Contradictions and Challenges

Brazil's cities are a result of a social structure, characterized by different life conditions and access to housing, services and urban equipment. The Brazilian housing crisis is characterized by the existence of irregular settlements with large populations, which usually have poor or no environmental sanitation. The most vulnerable populatons live in the most precarious areas, which are sometimes located in the marginal zones of streams and slopes, which puts families in life-threatening situations. In Brasil, despite the recent improvement in social housing production and programs designed to upgrade and give entitlement to the slums, there is already a huge deficit.

Housing is a basic human necessity. All of us depend on a shelter to survive the elements of the natural environment. As far as urban social life is concerned, our homes are linked to transportation, water, electricity and the communications network, which represent accesses to city life.

Our nation recognized the United Nations (UN) Resolution that affirms that housing is a human right, which include their affordability, tenure, salubrity, safety and security. However, since we live in a capitalist and profoundly unequal society, a house is also a commodity, that you should pay for rent or buy. Thus, part of the poorer population lives in illegal, and generally precarious, settlements. In Brazil they are classified and named as: slums (*favela, invasão*), informal settlements or informal parceling (*loteamento irregular*) and social housing units (*conjunto habitacional popular*).

Generally, these housing units are self-made, or is extended without adequate materials or technical assistance.

In developing countries, on the periphery of the capitalist system, the informal market is very large (work and employment, food commerce, industrial electronic devices, goods transportation and housing). But the market rules are not different for the rich or poor, neither for legal or illegal housing. The social stratification is reproduced on different housing markets' stratification. The informal market is also guided by the aim of getting some differential values that can bring rewards to the out by law owner (mainly that of the houses or huts built by their owners), who is, at the same time, a trespasser. There are also part of dwellings and rooms for rent as a consequence of the illegality, these houses are devaluated, when compared with a similar building and location in a legal situation.

Most of the population that lives in the slums does not have social and economic conditions (income, employment, living prospects). So, these dwellers cannot afford to pay for a house, either in terms of private real estate or public house-buying programmes that offer low instalments. It follows that slum dwellers are those social groups that should be the focus of state policies in order to create conditions for social and economic development.

It is the duty of the State to provide, by means of management of urban space and facilities, conditions for the social and economic integration of these excluded groups.

Water in the Urban Space

The presence of the water in the urban space is very important and it has different functions and perceptions. Without water the urban system (with its social and economic dimensions) could not exist. What happens upstream influences downstream – the quality and quantity of the flows. The rivers, creeks, springs and wetlands are the networks of the urban watershed. They are important for environmental comfort in terms of humidity and temperature. They have unique landscape qualities – ports, waterfronts, green and leisure areas. Generally, a city's sources of drinking water are found on the outskirts of the urban body. The hydrographic system defines – due to the topography – the drainage and sewage systems, which make healthy and sanitary conditions possible, despite the concentration of the population.

Managing the runoff is a challenge for a contemporary city due to the remarkable quantity of impervious areas, which enlarge the amount of rainwater that drains to pipes and reaches the creeks and rivers. The traditional urban infrastructure – pipes, gutters, ditches – have been constructed in order to replace small creeks and streams. The evaluation of some blocks in the Brazilian city of Campinas (undertaken by PUC Campinas students – Architecture and Planning) shows that the mean amount of pervious area in lots from medium class boroughs is 8 to 5% of the total area of the lot. The quality of runoff water varies depending on precipitation duration and also on the quality of the public services.

In Brazil, as is true in other developing countries, large parts of its cities do not have a sewage system. The collection system is incomplete and inefficient –only existing in some sections of the streets and only serving some dwellings. And even the sewage that is collected is not treated before being dropped into the creeks and rivers. Thus, citizens perceive the cities' hydrographic urban systems as dirty, ugly and evil. Mainly in the summer – the wet season in Brazil – flooding may occur. In the cities, the rainwater mixes with the contaminated water, enlarging the sanitary risks in the urban areas.

On July 28th, 2010, the General assembly adopted a resolution to recognize access to clean water and sanitation as a human right, by a recorded vote of 122 countries in favor, none against and 41 abstentions. Sanitary, security and affordability are principles of water access. However, there are several conflicts of interest due to the capitalist vision of tring to turn water into a commodity that is "produced" and sold by private companies to consumers. Considering water as a commodity means that everybody should have to pay to have access to it. Large numbers of countries that vote in the UN see the modernization of public sanitation, for the welfare of residents, as an opportunity to implement the "idea" that the water provided should be priced and bought. As a commodity, provided by the Market – private enterprises with financial resources by the State – water could be delivered to the consumer (no longer a citizen) "with adequate technical controls".

Water is basic a necessity for every human being. There is no life without water, and the scarcity of water is related to poor health conditions. Public health and welfare depend on safe and clean water. In Brazil, water is considered a public recourse. The sanitation systems should be provided and controlled by the municipalities.

Therefore, in Brazil's society, there are inequalities in living conditions and inequalities in the distribution of investment across the geographic space, both on regional and local scales – facilities, treatment and disinfection of the drinking water, treatment of sewage, garbage collection etc. In this context, in Brazil and similar countries there are large social groups that cannot afford to pay for water or other basic and social necessities. These people may live in poor areas, where such facilities are provided, or even the implementations of those facilities are not planned at all.

Water is also a strategic resource for several industrial sectors, mainly the food industry, agribusiness, chemical and pharmaceutical industries. The consumption, pollution and contamination, generated by those economic activities, create the scarcity of clean water for human use.

The richest and more educated citizens feel uncertain about the purity of the water, and are also pushed by industries advertising soft drinks and beverages. This sector has been buying water springs and developed a very extensive business – the bottled drinking water industry. The use of Poli (Tereftalate Ethilen) – PET – for packaging water and other liquids has a huge environmental impact, which is due it is an endocrine disrupter. The presence of PET bottles in our creeks and rivers, due to the poor garbage collection, also obstructs drainage pipes and increases the risk of flooding.

Housing and Water in the Urban Environment: Integrating Problems and Solutions

The main causes of the deterioration of water sources are: water pollution by domestic and industrial sewage, at higher levels that the self purifying capacity of rivers and excessive water removal, irrigation and industry production, compromising the human provision. Organic pollution added to the other contaminants present in pharmaceutical and personal care products is putting Brazil's society at risk.

Since 2003, the Federal Government has been raising financial resources to support projects designed to provide to better housing and living conditions, but mainly to build the sanitary systems' compounds. Brazil's society has been in a good position in terms of economic activities: growing consumption, high employment. But there is a historical lack of affordable and adequate housing for the general public, which is related to urban and architectural aspects.

In Brazil, growing investment for housing projects aimed at the "medium classes" accelerates the housing and urban development market sector. The best private land available in the inner-city was sold to those developments, and thus, urban land prices have increased. Thus, there is an "alternative" to the slums and irregular settlements for the poorer citizens. Because of the densification of those areas, it is currently necessary to remove 20% to 30% of dwellings to properly develop projects to upgrade the slums.

Climate change and urban environment

The main forecasts as a result of climate changes are: increase in extreme weather events such as tornadoes, thunderstorms, windstorms, heat waves, blizzards; increase in sea levels, a problematic issue due to the location of the populated areas; droughts, that may cause loss of agricultural areas, forest depletion, desertification, migration; greater incidence of vector-borne diseases (typhoid fever, dengue fever, cholera, schistosomiasis etc); higher incidence of circulatory diseases; lower incidence of respiratory diseases; increase in industrial risks: explosions at atomic power plants, petrochemical plants, pipelines. The population living in larger coastal cities are already very vulnerable to these risks (Nobre, 2010; Bueno, 2008).

As can be noted, these problems are related to water (scarcity or flooding) and at risk or inadequate human settlements. Thus, it is necessary to face these future indicators with social and political necessities.

A precautionary principle confronts the market-orientated society. Capitalist activity depends on risks and few regulations. On the other hand, in order to promote mitigation and adaptation, the State sector should be large and strong. It is necessary to develop others research projects, regulations and improve control. Mankind is not informed or conscious: scientific organizations, civil organizations and public structures must inform and promote participatory processes.

And About the future?

Nowadays it is difficult to think about this and act towards reaching a desirable future (as individuals, social groups or participants in state management). Generally, we (and our proposals) are limited by the Brazil's present problems – poverty and vulnerable people, unemployment, traffic congestion, contamination, risk areas.

Trying only to solve the past will not make more sustainable cities in the future. Creating a better future necessitates constructing a vision for the future, with clear goals. It is necessary to acquire some prediction capacity. Coping with the present problems and achieving desirable goals requires developing management – a necessary action to improve the capacity to provide.

Sustainable city – is it possible?

As this article has attempted to demonstrate, the city can be interpreted as an open ecosystem, whose metabolism we can understand and improve, since we recognize the political, economic and social powers that participate in it.

Henri Acselrad (2004) names sustainability as a "notion," not a concept. He points out that since the actual situation is unsustainable, one should consider actions towards a more sustainable situation, as compared with the given situation. The notion of sustainability should be a wonderful tool for designing a vision of the future, to measure or perceive indicators for promoting change in the urban space. But the marketing of the capitalist system has been trying to become the word, a miscellaneous set of concepts, manipulated as per the speaker's desires. Implemented by private sector, urban sustainability generally means more expensive land, housing, infrastructure and exclusive technologies, instead of more affordable, comprehensive and universal solutions. Economic development promotes more injustice. The poorer sections of society are living in contaminated and risky areas, which are the negative impacts of processing tons of raw material, garbage and sewage.

Ignacy Sachs created the noun ecodevelopment in 1970's. In earlier studies, Sachs proposed the dimensions of sustainability – social, economic, ecological, cultural, spatial, ethical and global and local politics.

Fritjop Capra, a researcher in Physics, was also an ecologist and an environmental scientist. He developed the proposal of the "ecological project," based on natural processes. Thus, energy cycles, nets and flows should be analyzed to develop a "Zero waste economy".

Rogers (1997) proposes that the city metabolism is managed in a circular direction rather than the linear metabolism that rules today, in which the city's population (during their economic and social activities) consumes energy and materials and exports a lot of its emissions to the outskirts (in solid, liquid or gaseous form).

Water and its hydrologic cycle are influenced by carbon and waste processing. Contamination is possible in each part and their connections. Thus, cleaning the environment is necessary to diminish the quantity of waste. Emissions must diminish in different points of the system: cargo transportation, use of packaging, maintaining the cleanness of public spaces etc. This process should be applied along the entire river watersheds, on both local and regional scales.

Guidelines

Urban sustainability can only work with more affordable, diverse and comprehensive solutions proposed to the human society as a whole.

Some principles or guidelines can be disseminated all over the world, such as: planning and management for the public and social values; dismantling over-consumption and waste economics and organize a production system for a Zero-emission economy; development of frugal innovation with the simplification of goods; social, communitarian and individual autonomy.

Some proposals have been undertaken in several cities in several countries, which are related to urban economy and urban design and management: development of Action plans for watersheds (regional and local scales), complementation of facilities, infrastructure and services) to improve urban areas and housing conditions; disseminating the culture of a more compact and sustainable city – control of urban sprawl, increase of popular and social housing in the inner city; rehabilitation of the central areas and vacant sites; recycling of waste to construct a new industrial chain; urban forestry – streets, roads, landscaping, public facilities, yards and gardens; dissemination of solar energy; radical reduction of solid waste; clean and extensive public transportation, and the increase of employment in activities related to sanitation, maintenance, health and social and elderly care.

References

ACSELRAD, H. Discursos da sustentabilidade urbana. *Revista Brasileira de Estudos Urbanos e Regionais*, ANPUR, n.1, p.79-90. May 1999. Available in: http://www.anpur.org.br/revista/rbeur/index.php/rbeur/article/view/27.

BUENO, L. M. M. Reflexões sobre o futuro da sustentabilidade urbana a partir de um enfoque socioambiental. *Cadernos Metrópole*, PUC-SP, São Paulo, n.19, 2008. Available in: http://revistas.pucsp.br/index.php/metropole/article/view/8712.

CAPRA, F. *Hidden Connections*: Integrating the Biological, Cognitive, and Social Dimensions of Life Into a Science of Sustainability. Nova York: Doubleday, 2002. See more in: http://www.fritjofcapra.net.

NOBRE C. et. al. *Vulnerabilidade das megacidades brasileiras às mudanças climáticas*: a Região Metropolitana de São Paulo. June 2010. Available in: http://www.inpe.br/noticias/arquivos/pdf/megacidades.pdf.

ROGERS, R. Cities for a small planet. London: Faber and Faber Limited, 1997.

SACHS, I. Estratégias de transição para o século XXI para pensar o desenvolvimento sustentável. São Paulo: Studio Nobel, 1993.

Sites

Fórum Brasileiro de Mudanças Climáticas. *www.forumclima.org.br*.

Instituto Brasileiro de Geografia e Estatística IBGE. *www.ibge.gov.br*.

Ministério das Cidades. *www.cidades.gov.br*.

Ministério de Ciência e Tecnologia. *www.mct.gov.br*.

Ministério do Meio Ambiente. *www.mma.gov.br*.

United States Census Bureau. *http://www.census.gov/population/international/*.

2

Ecolinguistics and Environmental Education in Brazil

José Marcelo Freitas de Luna[1]

Introduction

My training as an applied linguist appears to be a determining factor in many experiences that I have had as a foreign language teacher in Brazil and an as a researcher of teaching and learning situations in various other countries. Studying and working in countries like the United States and England, at universities with students and teachers from all corners of the globe, were, in fact, unforgettable and inspiring practices in my motivation to apply linguistics in multicultural environments.

Thus motivated, I set myself the goal, for the 2011 to 2012 period, of taking part in Top China Santander. The period leading up to the 2012 event is justified by the preparation I underwent, i.e. becoming familiar with the subject of the Program. I shall explain: although I considered urban "sustainability and planning" to be areas that are not only important but also interesting, my knowledge of them for teaching practice, in 2011, did not seem to be sufficient.

1 University of Vale do Itajaí, Santa Catarina, Brazil. E-mail: *mluna@univali.br*.

In fact, I was not yet ready to take part in the Program. I needed to take advantage of my training as a linguist, applying it to the important, interesting and mutidisciplinary area of sustainability. Thus, over a period of one year, I dedicated myself to reading essays on ecolinguistics and environmental education.

I consider this preparation to have been very important for all stages in the process. I refer, here, to the selection process at my university of origin, and the work that I would carry out at Peking University. In order to be selected, I had an interview, and gave a simulated lesson, both in English, with specialists in the area of sustainability and management of the Top China Program. Of course, while in Beijing, I was in direct contact with both Brazilian and Chinese teachers and students in this area.

It goes without saying that the first, and most insistent question I was asked during the interview was whether I, as a linguist, would be suitable for the Program. What does or can a language teacher do on the subject of sustainability?

To this question, and its various paraphrases, I replied environmental education has its theoretical premises, and maintains various relationships with general education. It is important to know that designing and developing sustainability activities are processes that are more successful when the communities involved are properly informed.

Objectively, I sought to justify my participation by appealing to the relationship between ecolinguistics and sustainability, which became not only an object of interaction during the interview and the simulated lesson, but a real display to students and teachers in China.

In this chapter, I present a report on a lesson I gave on the subject of Ecolinguistics and Environmental Education in Brazil during the Top China Santander 2012 Program at Peking University. In order to give readers of this book a knowledge of the abovementioned experience, the text is divided into four sections. The first describes how a rapport was initially established with the group of students, through warming up activities. In the two sections that follow, the basic concepts of the subject are outlined, showing their application to the teaching context of an Indigenous community school in Brazil. The chapter concludes with a section that brings together information on the expected and necessary interaction between the Chinese and

Brazilian participants on the theme of sustainability, from a perspective of ecolinguistics and environmental education.

Warming up

As appropriate for a group of young students, the subject was introduced using motivational activities in order to check their prior knowledge on the subject. A timed quiz was given, with carefully chosen prizes. Some typical quiz questions were:

- What's the word for "fish" in Chinese? And in Portuguese? In Danish, Romanian, Armenian?
- How many languages are spoken in the world today?
- Do languages die out?
- Do all languages have a written form?
- How many languages can be said to be safe from extinction?
- What's the word for "fish" in Tuvan, spoken by 200,000 inhabitants in Siberia and in Mongolia?

(Adapted from the text Dicionários Falantes, Revista Língua Portuguesa, year 7, n.80, June 2012.)

As I had predicted, the answers to these and other questions were not clear-cut. However, all the students gave their impressions and opinions, and as I had hoped, they were interested in learning more about the linguistic phenomenon, and in particular, the relationship between the language and culture of a people.

Once the group had demonstrated its motivation, I was able to present, in a more specific way, the aim of the lesson, which was: to provide the students with knowledge related to ecolinguistics and environmental education in Brazil, with a particular emphasis on indigenous language and education.

- I also presented the program to be developed:
- The concepts of Ecolinguistics and Environmental Education;
- The role of Higher Education Institutions;
- Environmental curriculum;
- The Infusion approach;

- Environmental teaching and learning activities;
- Some remarks;
- Assessment.

Presenting the cornerstones

The theme of linguistic education of the Indigenous people in Brazil was revealed to the students as one of the applications of the science of language. With this additional motivation, I was able to present the concepts that would be dealt with more specifically.

Language was presented as an activity; an ongoing work of subjects, with their specific historical, social and cultural contexts, and who, through this work or activity, organize, interpret and give shape to their experiences, and to the reality in which they live. Mother tongue, in turn, was understood as being situated in the scope of language, in the form of a system of intra/interpersonal and intra/intercultural communication, shared and used by members of one or more communities, through individual, geographical and social variations. Thus, linguistics was revealed to the students as the science that combines the different activities of systematic study of verbal language and developed natural languages.

Building on this theoretical framework, I was able to present and situate two further concepts. Ecolinguistics has been classed as a new branch of linguistics that investigates the role of language in the development and possible solution of ecological and environmental problems. As such, it aims to preserve the relationship between linguistic and biological diversity.

To check the level of learning of the concepts presented thus far, and maintain the students' motivation, I presented some case studies to illustrate the relationship between mother tongue and the environment, a fact that led to the consideration that much of what humans know about nature is encoded only in oral languages. Indigenous groups that have interacted closely with the natural world for thousands of years often have profound insights into local lands, plants, animals, and ecosystems.

Observing the students' learning, I returned to the framework of conceptual definition, formally presenting environmental education as a process of recognizing values and clarifying concepts, in order to develop the skills and attitudes necessary to understand and appreciate the interrelatedness among man, his culture, and his biophysical surroundings.

Bridging in

The practical part of the lesson was returned to, and continued until the end, with the showing of an edited film, lasting thirteen minutes, with subtitles in English. The film in question was the documentary of the series *Caminhos da Escola – Educação Indígena*, a TV Escola production from the Ministry of Education.

The film showed the theoretical and methodological approach of Brazilian educators in Indigenous communities of the Amazon. It was set in the town of São Gabriel da Cachoeira, located more than 800 kilometers from the capital Manaus. It is a region that has more than 400 communities and 35 thousand Indians belonging to 23 different groups. The school described in the film, and in the lesson, as an example of the use of the linguistic education methodology, was the Tukano Ye Pa Mahsa. The school divides students according to the different years of basic education, and also has a technical teaching group in sustainable indigenous development.

In the reports of the Indigenous students interviewed in the film, there was a marked difference between what they called conventional school, and the school of "our" community. For them, conventional education does not explore, or give value to the local culture, i.e. the habits, values, rituals and languages of the indigenous community. This fact, according to the community leaders, is causing students to drop out of school and migrate from the rural areas towards the cities.

With the development of the school system in indigenous communities, characterized by a curriculum that focuses on disciplines that relate to the students' actual needs and expectations, it is reported that pupils are staying in school more, and that community schools are being maintained. Another curricular component that was deemed positive by the Indigenous students was the use of their native languages as the primary language of teaching, alongside Portuguese. The various languages are, therefore, given importance, serving, naturally and consequently, as a means of expression and preservation of the local culture, including the environment.

In order to illustrate a common pedagogical practice at the Tukano Ye Pa Mahsa school, the documentary shows the production of *aturá*, a basket made from cipó vine, using weaving techniques that are disappearing. Through this teaching strategy, the functionality of the school curriculum, and the infusion of linguistic and cultural knowledge with the objectives of sustainability, become evident.

Encouraging the Students to Interact

The reactions of the students – both Chinese and Brazilian – to the content of the documentary, was surprising, in a positive way, in terms of what happens in Brazil with regard to linguistic and environmental education.

Presenting an example of a positive and successful pedalinguistic and environmental practice was, I confess, a special teaching and learning strategy that was specifically planned for the group. This is because the image and the representation that foreigners have of Brazil is not usually based on scientific fact. This misconception leads to stereotypes, with their consequent effects on negative attitudes in relation to what the country is or appears to be.

Aware of my commitment to ensuring scientific rigor, and at the same time, promoting the science and educational practices of Brazil, I emphasize that I opted strategically to show a present-day experience, in order to then refer to the Brazilian past.

In regard to linguistic rights, the period referred to as colonial in Brazil is marked by the application of elements of segregation with those of assimilation. The overall result was the fragmentation of the indigenous people and the death of indigenous languages. It is possible to date the beginning of the field of EE in Brazil back to the 1970s, as a response to the international debate generated by the Stockholm Conference (UN 1972). It is only since the 1980s that Environmental Education has clearly found its own place.

Under this climate of opinion, the Brazilian Constitution was passed in 1988. It guarantees the study of indigenous languages and the provision of indigenous communities through education. This achievement is seen by Brazilian society as a real benefit for environmental understanding and conservation efforts.

One of the direct consequences for schools in general is the inclusion of Environmental Education as a necessary item in important documents such as Agenda 21 and the Curriculum Parameters defined by Ministry of Education in 1998.

After summarizing some facts, past and present, related to linguistic and ecolinguistic education in Brazil, I then created the most objective and consistent condition possible for this vitally important interaction between Brazilian and Chinese students. I refer, here, to the intercultural discussion on how, in the universities of the respective countries, professionals can be trained to carry out their work in a way that is focused on sustainability.

Objectively, the "university" institution was, in ideal terms, defined by universality – its commitment to open-mindedness and openness to the world. When students graduate with the ability to think critically and act responsibly, then higher education will come closer to meeting its historic missions, not only in regards to advancing knowledge, but also to contributing to stable, more equitable democratic societies.

To support the discussion, I referred to the process of curricular infusion; a process through which environmental elements are incorporated into the program of a discipline; a bibliographic review of environmental issues is used in the readings and assignments, research projects on the field are carried out, and events promoting discussions about environmental themes are supported by the institutions.

I added that all too often, the common assumption is that only certain classes lend themselves well to environmental education. But I emphasized that all courses can be infused, through activities that can be characterized as stand-alone environmental courses; dedicated class sessions; or finding natural points of entry.

Although I had planned an essay-writing activity, when I met the group and saw the need for more cultural interaction between the students, I decided to develop a discussion forum instead. In pairs, they discussed and reported on Environmental Education activities applied to their courses in China and in Brazil.

The focus was on what components of sustainability were infused into the curricular disciplines of each course, if it was done at all. In general, the students were of the opinion that their courses directly or indirectly addressed the environmental theme, and that they were therefore aware of the importance of studying and acting in a professional manner, from the point of view of sustainability.

In relation to the expected and desirable comparison between Brazil and China, the reports of the groups indicate stages in the formulating and carrying out of educational objectives that are very similar in both countries. Or to put it another way, both the Brazilian and Chinese IHE have benefited from a climate of favorable opinion towards education and sustainability, infusing into their curricula concepts and practices that can be referred to as environmental education and ecolinguistics, particularly those that promote the preservation of languages and cultures of the peoples of Brazil and China.

References

ARBAT, E.; GELI, A. M. (Eds.). *Ambientalización curricular de los estudios superiores*. 1 – Aspectos Ambientales de les universidades. Girona: Universitat de Girona – Red ACES, 2002.

BRITISH Petroleum. Ten Years of Environmental Education in China. Press release, June 30th, 2009.

CARVALHO, L. M.; CAVALARI; R. M. F.; SANTANA; L. C. O processo de ambientalização curricular da UNESP – Câmpus de Rio Claro. Diagnóstico e perspectivas. In: GELI, A. M.; JUNYENT, M.; SÁNCHEZ, S. *Ambientalización curricular de los estudios superiores*. 3 – Diagnóstico de la Ambientalización Curricular de los Estudios Superiores. Girona: Universitat de Girona – Red ACES, 2003. p.131-65.

DÉCADA das Nações Unidas da Educação para o Desenvolvimento Sustentável – 2005-2014. Documento final. Plano Internacional de implementação. Brasília: Unesco, 2005.

FÓRUM Internacional de Ongs e Movimentos Sociais. *Tratado das ONGs*: aprovados no Fórum Internacional de Organizações Não Governamentais e Movimentos Sociais, no Âmbito do Fórum Global – ECO 92. Rio de Janeiro: Fórum, 1992.

INTERNATIONAL Working Meeting on Environmental Education in the School Curriculum. Final Report, September 1970. Gland, Switzerland: IUCN, 1970.

YEN, N. E. Education is the basis of environmental protection. *Environmental Science*, n.4, p.5-7, 1993.

WORLD Wildlife Fund for Nature (WWF). WWF in China. Beijing, 2008.

ZHOU, Z.Y. Deepening educational changes, strengthening environmental education. Speech delivered at the National Environmental Education Working Congress, November, 1992.

3

Water Resources in China: situation, measures and actions

Jiahui Shao[1]

Introduction

Water is one of our most critical resources. Up to 80 nations experience chronic fresh water shortages, which affect over 2 billion people, including an estimated 400 million children. The World Bank predicts that two-thirds of the world's population will suffer from lack of clean and safe drinking water by 2025. An article that appeared in *Fortune* magazine in 2000 predicted that "Water promises to be to the 21st century what oil was to the 20th century: the precious commodity that determines the wealth of nations".

In China, water is a highly precious resource and even becomes a limiting factor that plays a critical role in terms of economical and socio-economic development in many parts of the country. As its economy and population continue to grow, China will have to face greater and greater pressures on water resources. This paper aims to summarize the characteristics of water resources in China, and

1 School of Environmental Science and Engineering. Shanghai Jiao Tong University, Shanghai, China. E-mail: *jhshao@sjtu.edu.cn*.

the measures and actions taken in China. A special project on water pollution (SPWP) and the South-to-north water diversion (SNWD) were highlighted.

Water Resources in China

According to recent national a water resources evaluation conducted by the Ministry of Water Resources of China, China's annual precipitation is 648 mm, which represents a total amount of annual precipitation of 6,080 billion m^3. Taking into account the duplication between surface water and ground water, the total amount of available water resources on an average year in China is around 2,770 billion m^3, which includes 2,670 billion m^3 of mean annual surface run-off and 810 billion m^3 of groundwater per annum (Zhang; Xia; Hu, 2009). Despite China's total amount of available water resource being ranked sixth largest in the world, its available water resources per capita is only 2200 m^3, which is about one quarter of the world average. The amount of water resources per capita is expected to decrease by 1,750 m^3 in 2030 when China's projected population peaks at 1.6 billion (Zhang; Xia; Hu, 2009). Table 1 summarizes the China's change in water resources from 1999 to 2011 (The Ministry of Water Resources of the People's Republic of China, MWR, 1999-2011).

Based on water resources data in China Water Resources Bulletin provided by the MWR, annual precipitation and available water resources have shown a decrease trend in recent years, especially in the north of China, the most prominent year being 2011. In 2011, the mean annual precipitation was 582.3 mm, 16.3% less than that of the previous year and 9.4% less than the average year. The total amount of water resources was 2,325.67 billion m^3, 16.1% less than the average year and the lowest since 1956. Growing population, rapid economical development, urbanization, wasteful water usage as well as lax environmental oversight have all affected water resources in China. The basic characteristics of China's water resources are summarized as follows.

Table 1. Water resources change in China from 1999 to 2011

Year	Annual precipitation (mm)		Total water resources	
	Amount (mm)	Variation (%)	Amount (billion m^3)	Variation (%)
1999	629.0	-1.6	2,819,60	2.7
2000	633.0	-0.9	2,770.10	0.0
2001	612.0	-4.3	2,686,80	-2.2
2002	660.0	3.2	2,825.50	2.9
2003	638.0	-0.7	2,746.00	-1.0
2004	601.0	-6.5	2,413.00	-12,9
2005	644.3	0.3	2,805.30	1.2
2006	610.8	-4.9	2,533.00	-8.6
2007	610.0	-5.1	2,525.50	-8.9
2008	654.8	1.9	2,743.40	-1.0
2009	591.1	-8.0	2,418.02	-12.7
2010	695.4	8.2	3,090.64	11.5
2011	582.3	-9.4	2,325.67	-16.1
1999-2011 average	627.8		2,669.43	
mean average year	648.0		2,770.00	

Very low water resources per capita and per ha

With a vast area of 9.6 million km^2, China has relatively abundant water resources, being ranked sixth largest in the world regarding this aspect. However, per capita availability of water resources is only 2200 m^3, one quarter of the world average. China is listed as one of the countries in the world with the most severe shortage of water. In addition, water resource availability per ha of cultivated land in China is 1391 m^3, which is only equivalent to 69% of the world average.

China's major water resources include 9 main river basins, which are as follows: (1) Songhua-Liao River Basin, (2) Hai-Luan River Basin, (3) Yellow River Basin, (4) Huai River Basin, (5) Inland River Basin, (6) Yangtze River Basin, (7) Pearl River Basin, (8) Southeast River Basin and (9) Southwest River Basin. Zones (1) to (5) are in northern China, and zones (6) to (9) are in southern China. Table 2 provides statistics regarding water resource availability per capita and per ha for the nine zones in China (Shen, 2004).

Highly uneven distribution across the country

Table 2 clearly shows that water resources are comparatively abundant in the south, but much less so in the north of China. China's highly uneven spatial distribution of water resources does not match with the distribution of population or agricultural land. As shown in Table 2, 80.9% of China's water resources are in southern China, which is only home to 53.6% of its population and 34.7% of its arable land. Per capita water resource availability for southern China is more than four times greater than that of northern China, and the per ha water availability for arable land in southern China is more than ten times greater than that of northern China. This geographic mismatch is particularly severe in Hai, Huai, and the Yellow river basins, which only have 7.6% of the nation's water resources yet cover 34.7% of the population and 39.4% of the arable land.

Rivers represent the major source of water for all the nine zones. There are numerous rivers with a total length of 0.42 million km, of which about 1500 rivers drain into a basin area of over 1000 km^2 each. The groundwater resource is only important on the plains, especially in northern China. It is negligible elsewhere in the country.

Table 2. Statistics of water resources availability per capita and per ha for nine river basins in China

	River Basin	Water available		National percentage		Water Avaliability per capita (to m²)		Water avaliability per ha (m³)
		percent (%)	1000 m³	Population	Arable Land	1997	2010	
Northern rivers	Inland	4.6	130.4	2.1	5.7	4,876	4,140	23,835
	Songhua-liao	6.9	192.2	9.6	20.2	1,646	1,501	9,900
	Hai-Luan	1.5	42.2	10.0	11.3	343	311	3,885
	Huai	3.4	96.1	16.2	15.2	487	440	6,555
	Yellow	2.7	74.4	8.5	12.9	707	621	6,000
	Nort Total	19.1	535.3	46.4	65.3	8,059	7,013	50,175
Southern rivers	Yangtzé	34.2	961.3	34.3	23.7	2,289	2,042	41,745
	Pearl	16.7	470.8	12.1	6.7	3,228	2,813	67,515
	Southeast	9.2	259.2	5.6	2.5	2,285	2,613	80,160
	Southwest	20.8	585.3	1.6	1.8	29,427	25,056	346,350
	South Total	80.9	2277	53.6	34.7	34,001	32,524	535,770
	Total	100	2812.3	100	100	42,060	39,537	585,945

Highly uneven distribution throughout the year and between years

As a result of China's monsoon climate, rainfall occurs unevenly over the year, leading to an uneven distribution of water resources throughout the year. Most precipitation occurs in four consecutive months from April to July in southern China, accounting for about 60% of the total surface runoff per year. In Northern China, over 80% of surface runoff occurs from June to September, and 90% in some places, especially in the Hai river basin. The water runoff gap between years is also significant. Ratios of the maximum to the minimum annual runoff are usually greater than five and even reach 10-20 in the north, falling within 2-3 for southern rivers (Jun; Chen, 2001). This variability causes alternating floods and droughts in river basins across the country (Xu; Dai, 2002).

Deteriorating water quality

According to the national surface water quality standards of China [GB3838-2002], ambient water quality is divided into five categories, which are as follows. (1) Grade I, mainly applicable to sources of water bodies and national nature preserves; (2) Grade II, mainly applicable to class A water source protection areas for centralized drinking water supply, sanctuaries for rare species of fish, and spawning grounds for fish and shrimp. (3) Grade III, mainly applicable to class B water source protection areas for centralized drinking water supply, sanctuaries for common species of fish, and swimming zones; (4) Grade IV, mainly applicable to water bodies for general industrial water supply and recreational waters in which there is no direct human contact with the water; (5) Grade V, mainly applicable to water bodies for agricultural water supply and for general landscaping requirements. This water quality classification system is based on purposes of use and protection. Grade I, II and III permit direct human contact and use as raw water for potable water systems. Grade IV is restricted to industrial use and recreational uses other than swimming. Grade V is restricted to irrigation.

Results from ongoing water quality monitoring show overall water quality in China to be poor. Generally, river water quality is better in the upper reaches than in the lower reaches, mainstreams have better water quality than tributaries, and urban streams are among the most polluted (Jun; Chen, 2001). Rivers in the south are generally cleaner than those in the north, which is because southern rivers generally have larger assimilative capacities that is due to the relatively more abundant runoff. Rivers and streams in economically more advanced regions in the east have poorer water quality than those in the less developed central and western territories. In 2011, the surface water across the country was graded as being slightly polluted (Ministry of Environmental Protection of the People's Republic of China, MEP, 2011). In the 469 river sections of the ten big water systems such as the Yangtze River, Yellow River, Pearl River, Songhua River, Huai River, Hai River, Liao River, rivers in Zhejiang Province and Fujian Province, rivers in southwestern China and inland rivers under the national monitoring program, 61.0% met Grade I~III national surface water quality standard, 25.3% met Grade IV~V standard and 13.7% failed to meet Grade V standard. The main pollution indicators were COD, BOD5 and total phosphorus (TP).

Water quality problems in lakes and reservoirs are primarily associated with eutrophication. In 2011, among the 26 major lakes (reservoirs) included in the national monitoring program, 42.3% met Grade I~III standard, 50.0% met Grade IV~V standard and 7.7% failed to meet Grade V standard. The main pollution indicators were TP and COD. 46.2% of the above 26 lakes (reservoirs) were under mesotrophic state, 46.1% were under slight eutrophication and 7.7% were under intermediate eutrophication (MEP, 2011).

Despite considerable efforts to clean these bodies, China's water pollution is still a major problem. Table 3 shows a summary of the amount of wastewater and major pollutants emitted into China's rivers based on the data from Report on the State of the Environment in China from 1999 to 2010 (MEP, 1999-2010).

Table 3 shows that the amount of total wastewater emitted has increased since 1999. The total amount of wastewater emitted in 2010 was 61.73 billion m^3, which is 53.9% more than the 40.11 billion m^3 in 1999. From 1999 to 2007, the total industrial wastewater volume increased and peaked in 2007 at 24.66 billion m^3, 24.9% more than the 19.73 billion m^3 in 1999, and began to decline to 23.75 billion m^3 in 2010. Improved and more stringent regulation on industrial wastewater, as well as improved industrial processes and an increase in the

Table 3. Wastewater emission and major pollutants in China's rivers

Year	Amount of wastewater (billion m^3)			COD (million m^2)			Amonia-nitrogen (million m^3)		
	Total	Industrial	Domestic	Total	Industrial	Domestic	Total	Industrial	Domestic
1999	40.11	19.73	20.38	13.889	6.917	6.972	na	na	na
2000	41.45	19.42	22.09	14.459	7.045	7.045	na	na	na
2001	43.29	20.26	23.03	14.048	6.075	7.973	1.252	0.413	0.839
2002	43.95	20.72	23.23	13.669	5.840	7.829	1.288	0.421	0.867
2003	46.00	21.24	24.76	13.336	5.119	8.217	1.297	0.404	0.893
2004	48.24	22.11	26.13	13.392	5.097	8.295	1.330	0.422	0.908
2005	52.45	24.31	28.14	14.142	5.548	8.594	1.498	0.525	0.973
2006	53.68	24.02	29.66	14.282	5.415	8.867	1.413	0.425	0.988
2007	55.68	24.66	31.02	13.818	5.111	8.708	1.323	0.341	0.983
2008	57.20	24.19	33.01	13.207	4.576	8.631	1.27	0.297	0.973
2009	58.92	23.44	35.48	12.775	4.397	8.378	1.226	0.273	0.953
2010	61.73	23.75	37.98	12.381	4.348	8.033	0.273	0.273	0.930

volume of wastewater treated, are the main reasons for this decrease in industrial wastewater volume in recent years. However, in the same period, there has been a much larger increase in total domestic wastewater volumes, rising from 20.38 billion m^3 in 1999 to 37.98 billion m^3 in 2010. This led to a larger increase in total wastewater discharge from 1999 to 2010, and is consistent with China's fast urbanization and economic growth, which has in turn provided its people with a improved quality of life. In spite of the fact that the total amount of wastewater discharge has increased, the level of chemical oxygen demand (COD) began to fall from 2000, the exceptions to this being in 2005 and 2006. The level of ammonia-nitrogen from wastewater discharge has followed the same pattern as the level of COD from 1999 to 2010. The decrease of COD and ammonia-nitrogen suggest that the water pollution problem may have been successfully addressed.

Measures and Actions Towards Sustainable Water Resouce Development in China

In 1994, the Central Government promulgated China's Agenda 21 (GPRC, 1994). Subsequently, China's Water Agenda 21 (MWR, 1998) was formulated in 1998 as a result of a number of studies conducted on water (Yang; Pang, 2006). Water Agenda 21 outlined policies for sustainable water resource development and listed key actions and projects. Priority was given to addressing water shortage and pollution issues, and to meeting the basic water needs of urban inhabitants, industry, agriculture, and ecosystems. In line with China's Water Agenda 21, action plans were formulated for the Yangtze river basin, Huai river basin, Yellow river basin, Hai and Luan river basins, Taihu Lake, Songhua and Liao river basins, and Pearl river basin. These plans have guided water resource development in each of these river basins. Legislation and government regulation have played a vital role in managing water resources and dealing with water problems in China. China has taken various measures and actions in response to the problems of its water resources, such as improving the legal water pollution management structure to be more comprehensive and simplified, carrying out integrated water resources management, emphasizing science and technology of water pollution control, improving water quality monitoring, law enforcement and capacity building, strengthening economic instruments and financing and enhancing environmental awareness and public participation. Two major projects in this regard are highlighted below.

Special project on water treatment (SPWP)

On Feb 9th, 2006, the State Council issued the "National Mid-Term and Long-Term Science and Technology Development Plan (2006-2020)," which provides guidelines for China's science and technology development for the next 15 years. The Project on the Prevention and Control of Water Pollution (hereinafter referred to as Special Project on Water Pollution, SPWP) is one of 16 national Major Special Research Projects listed in the plan. The SPWP's aim is to provide strong scientific support for China's water pollution control and treatment, as well as to reach the goal of a 10% reduction of COD and controlling the total emission of major pollutants at the end of 11th five-year plan. This project focuses on treating the entire river basins instead of the conventional approach of end-pipe treatment in order to guarantee safe drinking water and improve the overall water environment. Under the unified leadership of the State Council, the SPWP is coordinated by the MEP (the Ministry of Environmental Protection) and the Ministry of Housing and Urban-Rural Development (MOHURD). It is the biggest investment in environmental protection since the founding of the People's Republic of China in 1949, which has an estimated budget of more than 30 billion Chinese yuan (around 4.7 billion US dollars) over 13 years. The SPWP is implemented in three phases. The milestone objectives of these three phases are: (1) to make breakthroughs in key technologies regarding "controlling the source to reduce discharge" in the first phase (2008-2010); (2) to make breakthroughs in key technologies regarding "reduction and reclamation" in the second phase (2011-2015); (3) to make breakthroughs by completing a set of key technologies on "integrated regulation" (2016-2020). The ultimate goal is to establish two technological support systems for routine water pollution forecasting and monitoring, in addition to water pollution control, and to create an integrated management platform for national water environment. During the first stage, the SPWP identified 238 research projects in 33 programs with six themes focusing on major river basins, such as the "three rivers (Huai River, Hai River and Lao River), three lakes (Tai Lake, Chao Lake and Dian Chi Lake), one big river (Songhua River) and one reservoir (Three Gorges)," 230 research projects in 32 programs werelaunched during the "11th Five-Year Plan" period. Six themes are (1) lake eutrophication control and reclamation, (2) river water environment integrated renovation, (3) city water pollution control and water environment integrated renovation, (4) safe drinking water ensurance, (5) water environment monitoring and forecasting and (6) water environment strategic policy and management.

Until now, under the leadership of the leading group of SPWP, this special project has taken full advantage of the government-oriented system with coordination, innovation and facilitated industrial structural adjustment with the help of research, development and integration of technologies; demonstration projects and application of research findings. It has driven the pollution control process of some regions and enterprises and made remarkable achievements. It has played an active role in accelerating the transformation of economic development mode (MEP, 2011).

South-to-north water diversion project (SNWD)

The uneven distribution of China's water resources between the water-scarce north and water-abundant south is forcing the Chinese government to seek measures to ensure sufficient water availability for people living in the northern regions. The "South-to-North Water Diversion Project" is designed to better utilize water resources in southern China. The project includes three routes. The Western Route Project (WRP), Middle Route Project (MRP) and Eastern Route Project (ERP) are designed to divert water from the upper, middle, and lower reaches of the Yangtze River respectively, the objective of which being to meet the developing requirements of Northwest and Northern China (SNWD Project Commission of the State Council website). The project is designed to divert 13.4 billion m^3/yr of water from the Yangtze to Beijing, with the project supplying many other cities along the route too. The plan foresees a total water transfer capacity of 44.8 billion m^3 by 2050 (The World Bank, 2006). The complete project is expected to cost $62 billion, which is more than twice as much as the Three Gorges Dam. Up to 2012, more than $34 billion has already been spent. The total expenditure is expected to increase. However, the project has created widespread controversy since its introduction, its construction on the Eastern Route officially began on December 27, 2002 and the central route began in 2004. One of a number of logistical challenges for the project facing is to clean water bodies at the intersections through which the canals will pass. Water pollution has delayed the entire project process. In 2012, it was reported that the eastern route of the South-to-North Water Diversion Project has made obvious progress regarding pollution control, and will achieve its ambitious goal of supplying clean water at some time next year to Shandong, as is cited in news from the SNWD official site.

Conclusions

Severe water shortage and water pollution affect China's water resources. Population growth, socio-economic development, urbanization, and improvements in living standards have increased water demand and pollution, making water resources the major obstacle to China realizing is objective of sustainable development. It is encouraging to see that China has responded using various measures and actions to solve water-related problems, and progress has been made towards sustainable water use. It is hoped that sustainable water use will be achieved through various effective measures, actions and integrated water resource management in China.

Acknowledgements

This research was supported by National Natural Science Foundation of China (31061140360).

References

GPRC (Government of the People's Republic of China). *China's Agenda 21*. Beijing, China: China Environment Press, 1994.

JUN, X.; CHEN, Y. Water Problems and Opportunities in the Hydrological Sciences in China. *Hydrological Sciences-Journal-des Sciences Hydrologiques*, special issue, v.46, n.6, p.907-21, 2001.

MEP (Ministry of Environmental Protection of the People's Republic of China). *Report on the State of the Environment in China*. 1999-2011.

MWR. *China's Water Agenda 21*. Beijing, China: China Water and Power Press, 1998.

MWR (The Ministry of Water Resources of the People's Republic of China). *China Water Resources Bulletin*. 1999-2011.

SHEN, D. The 2002 Water Law: its Impacts on River Basin Management in China. *Water Policy*, 6, p.345-64, 2004.

SNWD Commission of the State Council. *South-to-North Water Diversion Project Commission of the State Council*. Available at: http://www.nsbd.gov.cn/zx/english/.

STATE Council. *National Mid-Term and Long-Term Science and Technology Development Plan* (2006-2020). 2006.

THE WORLD BANK. *Water Quality Management*: Policy and Institutional Considerations. Sept. 2006. Available at: http://www.worldbank.org/eapenvironment.

XU, Q.; DAI, D. *Study on the countermeasures against flooding in China*. Beijing, China: China Water and Power Press, 2002.

YANG, X.; PANG, J. Implementing China's "Water Agenda 21". *Front Ecol. Environ*, v.4, n.7, p.362-8, 2006.

ZHANG, L.; XIA, J.; HU, Z. Situation and problem analysis of water resources security in China (in Chinese). *Resources and environment in the Yangtze basin*, v.18, n.2, p.116-20, 2009.

4

Urban planning in Brazil

Ítalo Stephan[1]

Introduction

First came the caches, and then came the villages, towns and cities. There is some evidence that the origin of the cities is traceable to about seven or eight thousand years ago. A village is the origin of the city because it was formed by a set of homes, an oratory, public roads and squares. The urban world has been totally dependent on the rural world for centuries. While the rural world was place of food production, the urban world was one of power (temporal and spiritual), commerce, manufacturing and social life. The growth of cities was limited in the beginning, because for each resident in the city the work of hundreds of farmers was necessary. The development of agricultural techniques made the growth of cities possible. Nowadays growth is no longer limited, which is due to the rise in the agricultural productivity of farmers, where one can easily feed a thousand people. The development of agricultural techniques and Increasing agricultural productivity has made it possible for huge numbers to migrate from

1 Federal University of Viçosa, Viçosa, Minas Gerais, Brazil. E-mail: *stephan@ufv.br*.

the rural areas into urban civilization. In 2007, fifty percent of people on the planet were living in towns, cities, metropolises and megacities with 20, 30 and even 35 million inhabitants. However, this growth is unequal. The urban context is, in almost all cities in the world, socially unfair and full of contrasts.

Eight three percent of the Brazilian population resided in urban areas in 2010, which means that 161.800.000 people were living in towns and cities. In Brazil, the phenomenon of urbanization has exacerbated the historical context of social exclusion, marginalization and violence.

Thus, the growth of Brazilian cities in 2010 shows many contrasts. In the cities, on the one hand, there are some islands of wealth, with all the amenities found in developed countries, on the other hand, many poor neighborhoods do not have an adequate urban infrastructure. These contrasts also encompass a poor system of education, a precarious system of transportation, as well as increased violence and health problems caused by poor sanitation.

The Brazilian Urbanization Process

Rural Brazil

Brazil was essentially agricultural for more than four centuries. Our products came from mining, extensive livestock, coffee and cocoa plantations and rubber extraction. Until the mid-twentieth century the rural population was greater than the urban.

The urban Brazil

For decades, smoke, noise and congestion were synonyms of development.

Places without these characteristics were considered backward, underdeveloped or were recognized as provinces.

Until the nineteenth century Brazilian cities were basically coastal cities, functioning as harbors and Government headquarters, which were mainly in Salvador (the first capital) and Rio de Janeiro (capital until 1960). These cities were the connection route of the world to the gold mines (mostly from Minas Gerais province). Urbanization followed the process of industrialization. This began with the incentive provided by President Getúlio Vargas in the 1930s. Industries were firstly installed in Rio de Janeiro, São Paulo and then in Belo Horizonte.

Figure 1. Sugar mill. Engraving of Rugendas.

Source: *http://pt.wikipedia.org/wiki/Engenho.*

The new capital Brasilia was idealized by architects Lúcio Costa and Oscar Niemeyer and helped to attract people to central Brazil, which was previously deserted by people. The city was founded in 1960.

In the 1970s, the urbanization process in Brazil was considered one of the most rapid and chaotic in the world. The first Impacts: migration to the metropolises meant prosperity wealth and social climbing. Considering that growth and development are different, the process of Brazilian urbanization with a high concentration of people and high concentration of economic activities required increasing investment in infrastructure and services. As this did not happen, cities and their suburbs have grown haphazardly.

Figure 2. Rio de Janeiro in XIX century. Engraving of Rugendas.

Source: *http://literaturabrasileira3.blogspot.com/2008/04/crnica-sculo-xix.html.*

Brazilian Cities: Issues

Peripheries of Brazilian cities: the chaotic growth

Between 1998 and 2008 the population of eight metropolitan regions (Rio de Janeiro, São Paulo, Belo Horizonte, Vitória, Porto Alegre, Curitiba, Recife and Salvador) jumped from 37 million to 42 million people. Simultaneously the rate of growth of the suburbs of these cities was 30%; six times higher than the growth rate of wealthier neighborhoods in these cities. In 1970, São Paulo had something like 70,000 squatters. This number was just over 1% of the population, however the problem hardly worried the rulers of the time. Three decades after, the population jumped from 70.000 to 2.6 million people – equivalent to more than double the number of residents of a city the size of Campinas.

Figure 3. Brasília.

By Ítalo Stephan.

Peripheries of Brazilian cities: the threats

The periphery of the 50 largest cities in Brazil is now home to more than 80 million inhabitants. Nowadays, is the periphery represents almost half of all Brazil's residents. Violence has also grown considerably during this time. Ten years ago the homicide rate in these regions was 30 per 100,000 inhabitants. Nowadays it is three times higher: 150 deaths per 100,000 inhabitants.

Peripheries of Brazilian cities: the mismatch

Migration would not be a big problem if cities were able to generate enough wealth to provide satisfactory living conditions for all who came. There is a contradiction. Municipalities are in the forefront of the problems, but in last place in the decision-making chain to solve them. There are some possible explanations: the existence of an anachronistic federal structure (at federal, state and municipal levels); the lack of a culture of planning, especially with regard to their physical and territorial issues.

Brazilian cities – Findings

The current theme is: a decrease in quality of life and growth of the informal city. The perspectives are: an accentuation of the current stage of social, economic and environmental degradation. There is a necessity to promote development based on the alternatives to reduce social inequalities; improve respect for the environment; and promote continuous generation of wealth.

Brazilian cities – the exception

Curitiba is a city that is well known all over the world. It is the capital of the state of Paraná (south region) and has a population of more than two million people. It is famous because of its tradition in urban planning (since 1965); by its integrated public transportation system; by its city planning institute and by its dedicated architect Jaime Lerner. Curitiba is known as an exception in Brazil, and became an example of successful urban policies whose experiences spread to other cities around the world.

Brazilian cities: the conquests

The history of urban reform began in the 1960's, and, after the dictatorial period, was reduced to a couple of articles in the federal constitution from 1988. Articles 182 and 183 established that cities with more than twenty thousand inhabitants were required to have master plans and required that the urban property meets its social function. Along the process of formulating the plan, citizens must define when, where and how urban property performs its social function. Citizens define when a property remains empty or underused. After these discussions they decided which tools and urban parameters would become part of the plans.

The Statute of the City – Law No.10.257/01 regulated these articles and established general guidelines for urban policy based on democratic management; urban planning; ordination and control of land use; the right to sustain cities and integrate urban and rural activities.

Figure 4. Viçosa, MG.

By Ítalo Stephan.

The Statute of the City establishes goals that aim to ensure everyone's right to live well, the preservation of the environment and the inclusion of the population in the discussion about the city's growth.

Brazilian cities – an achievement

The Statute of the City also requires participation in three ways. The first is guaranteed through public hearings and debates with the participation of the population and associations that represent various segments of the community. The second is using advertising on the documents and information produced. The third way is guaranteeing access to the documents and information produced by any interested party.

Figure 5. Public hearing for the Master Plan of Guaxupé, MG.

By Ítalo Stephan.

A master plan became mandatory for cities that:

- have more than 20,000 inhabitants;
- are members of metropolitan regions and urban conglomerations;
- are members of special tourist interest areas;
- are inserted in the area of influence of developments or activities with significant environmental impact in the regional or national domain.

According the Constitution and the Statute of the City, having a Master Plan has become mandatory for nearly thirty percent of the 5,565 Brazilian municipalities. The Master Plan, more than a statutory requirement, is a means, an opportunity for citizens to rethink the city where they live and work. It aims to give order to the city and countryside and the development of the municipality. It is a city ordinance created with the participation of the whole society and is approved by Town Hall. According to the Statute of the City, a master plan must have:

- Contents;
- Statement of reasons;
- Definition of the social function of property (includes the definition of what is unused and empty terrain urban);
- Guidelines for public policies such as social housing and urban mobility;
- Tools for sustainable urban development and improvements to reach a better social equilibrium;
- Forms of popular participation (public queries, public debates, public hearings);
- Rules for land use and land parceling;
- System for planning and monitoring.

The statute sets instruments for urban development, such as preemptive rights; tools to control land use and its occupation; the created soil,[2] transfer of development rights – TDR,[3] impact study on the neighborhood[4] and the contribution for Improvement.[5]

---

2 A form of separating the right to property from the right to build. It is possible to design a previous potential increase above the zoning in force for a certain area of the city, granted for high cost (which are the cases of Porto Alegre and Curitiba), or according to the City Statute, in which the created soil is established above a basic single coefficient for all urban area or differentiated for specific areas within the urban area.

3 Public regulatory instrument for the right to build, by which the owner of urban estate, either private or public, can be entitled to practice the right to build elsewhere or sell this right, provided that the purposes are explained (implantation of urban and communitarian equipment, preservation of estates of cultural interest, or for land regularization, social housing). Instrument used in Curitiba since the 1960s. See Art. 35 of the City Statute.

4 Instrument aimed at the democratization of the decision making system related to ventures.

5 Redistributive taxation, payable by owners benefitted by the execution of public investment, "a form of compensation for the surplus-value created at the expense of collective investment, which was individually appropriated".

Other planning tools:

- Subdivision, building or compulsory use. Specific municipal law for areas included in the master plan can determine the subdivision, or compulsory use of unbuilt urban land, under-used or unused urban land, and should establish conditions and deadlines for the implementation of the referred to obligations.
- Property taxes (IPTU) that are progressive over time: in case of noncompliance with the conditions of sub-division, building or compulsory use, the municipality can proceed with applying taxes above the IPTU that are progressive over time, through increasing the tax rate over five consecutive years.
- Expropriation with payment in bonds: five years after the collection of progressive tax on property, if the owner has not complied with the obligation to divide, build or use the property, the municipality can move toward a condemnation with payment of government debt bonds. However, after seven years of the statute, most planning surveys have shown that those urban planning tools are rarely detailed in plans, mainly in a self-applicable form, and are almost never applied.

Surveys regarding the planning of small towns also show other issues. There is a strict dependence on the political interest of the mayors. If they are not interested, or think that they could have disadvantages, they do not give the plan any consideration. There is not even a small planning agency and city officials do not have planning skills, therefore significant parts of the plans will never be applied. Some actions, constructions and projects included in plans are not tied with the budget and are sentenced never to be undertaken.

Findings

Brazilian towns, cities and metropolises are still growing along with many problems.

There is a lack of a good housing policy, there are many environmental issues and most of these towns and cities still do not have their basic sanitation issues even partially resolved. The urban sprawl must be stopped at the same time, it is essential to find ways to reduce the high cost of land.

Figure 6. Social housing program undertaken in Brazil.

By Ítalo Stephan.

For small towns it is urgent to advance the simplification of these plans and the planning process. It is also important to move towards integration and regional planning.

There is still a long way of learning. Politicians must be better chosen, municipal technicians must be better prepared, people must be more participative, and must fight for their rights

The Statute of the City is still mainly an important theoretical advancement. After ten years that the Statute of the City has been in existance, there are more than 1,700 master plans in Brazilian cities, in the 27 states. Soon all the municipalities, even the smallest towns, will be required to prepare their master plans.

There are some studies that evaluate plans, and quite a few that evaluate the application of these plans. It is important to do this. Some of the first results are very inconclusive while some are disappointing. Good and bad experiences must be shown. Initial surveys show few changes, however these are over a short period of time. Public participation is still insufficient and unsatisfactory. It is necessary to draw on these lessons. The challenges of effective participation, good management and urban tools application must prevail.

Acknowledgements

I would like to thank the Santander Bank for the great opportunity to visit a small part of China. I would also like to thanks the organizers of an impressive scheme of logistics in Brazil and mainly in Shanghai and Beijing. With thanks to the Chinese people who received us with enormous kindness and who became great friends.

References

BRASIL. Lei 10.257, de 10 de julho de 2001. Estatuto da Cidade. Brasília: Câmara dos Deputados, 2001.

CYMBALISTA, R. Instrumentos de planejamento e gestão da política urbana: um bom momento para avaliação. In: BUENO, L. M. de M.; CYMBALISTA, R. (Orgs.). Planos diretores municipais: novos conceitos de planejamento territorial. São Paulo: Annablume, 2007. p.25-32.

MARICATO, E. *Brasil, cidades*: alternativas para a crise urbana. Petrópolis: Vozes, 2001.

SAULE JÚNIOR, N. Do plano diretor. In: MATTOS, L. P. (Org.) *Estatuto da cidade comentado*: lei n.10.257, de 10 de julho de 2001. Belo Horizonte: Melhoramentos, 2002. p.255-94.

SOUZA, M. L. de. Mudar a cidade: uma introdução crítica ao planejamento e à gestão urbanos. Rio de Janeiro: Bertrand Brasil, 2002.

STEPHAN, Í. I. C. *A aplicação dos planos diretores e leis de controle do uso e ocupação do solo em cidades de médio porte demográfico, em Minas Gerais, no período 1988-1998*. São Paulo,

2005. 381f. Tese (Doutorado em Arquitetura e Urbanismo) – Faculdade de Arquitetura e Urbanismo, Universidade de São Paulo.

STEPHAN, Í. I. C. Planos Diretores em Minas Gerais: vinte anos de exigência constitucional. *Revista Risco*, São Carlos, n.10, p.46-56, 2009. Available at: http://arquitetura.eesc.usp.br/revista_risco/Risco10-pdf/02_art05_risco10.pdf. Access in: March 1st, 2011.

VILLAÇA, F. Uma contribuição para a história do planejamento urbano no Brasil. In: CSABA, D.; Schiffer, S. R. *O processo de urbanização no Brasil*. São Paulo: FUPAM/EDUSP, 1999. p.169-243.

5

Sustainability, Development and Well-being at Work

Inês Monteiro[1]

Introduction

Discussions on sustainability and development have increased over the last decade, from the beginning of the 21st century, both in developed and developing countries.

An important aspect to be noted is regarding the ethic that pervades development nowadays, both in respect to companies and workers.

Based on a governmental perspective, in relation to ethical issues, it is crucial to eradicate corruption; respect the environment; and develop the country, the aim of which is to reduce social inequality and provide access to goods and services in different regions, particularly in countries with large territories, such Brazil and China.

Amaral (2011, p.54) reports that "Regarding the scope of the law, the central government has also sought to improve its labor laws to meet the interests of workers and to ensure some rights, even though it takes a while to be required before the federal courts of the country". Naisbitt (2010) apud Amaral (2011) affirm that "Since January

1 University of Campinas, Campinas – SP, Brazil. E-mail: *inesmon@fcm.unicamp.br*.

2008 the legal rights of China's working class have been strengthened by a new labor law. Now more than 800,000 workers can demand a working contract, and their demand is backed by penalties that can be imposed. It will take a while until the workers really dare to do this, but it is a further advancement of individual rights".

From the perspective of the companies, there is respect for the laws of all the countries in which they operate; attention to the production process; respect for the known rules in developed countries, although they have not yet been implemented in those which the rules work; in addition to guaranteeing workers with suitable working conditions and appropriate and fair remuneration.

In general, having a choice of whether to work in extremely unfavorable conditions is not really a reality for workers, because there are often very few job opportunities in the region.

It is pertinent, in this context, to remind ourselves of Wooding and Levinstein's question (1999, p.143): "what do workers or people from disadvantaged social class or impoverished countries do when faced with the choice – their work or their health?"

Another aspect worth noting is the discussion on the green economy. The European Commission proposed, for the 2020 Strategy, that governments should seek three objectives: "to be sustainable, intelligent and have inclusive growth, which includes promoting competitive low-carbon, resource-efficient and greener economy". Another aspect to be highlighted is the jobs to be created in sectors related to the environment and the new economic model, which may require certain skills and competencies (Greening, 2011, p.25).

Education also plays a relevant role in the search for sustainability, especially in the training of engineers.

Brazil and China: socio-demographic data

Brazil is a federal republic with 26 States and a Federal District. Its territory covers 8,515,692.7 km^2 and had a population of 190,755.799 inhabitants in 2010, and had 5565 municipalities, 15 of which had populations above one million inhabitants. The gender ratio was 96 men for every 100 women. There were 67.5 million residences at this time, with an average of 3.3 people in each particular residence. The urbanization rate in Brazil was 84.4% (IBGE – Brazilian Institute of Geography and Statistics, 2011). In 2009, the average fertility rate in Brazil was 1.94 children per woman (IBGE, 2010).

China is the most populous country in the world, with 1.3 billion people, and a life expectancy of 73.5 years. China is currently undergoing a rapid urbanization process, with over 200 million people in the process of migration from rural to urban areas. The fertility rate is 1.8, this rate being subsequent to the family planning program that began in the 1970s, according to information provided at the 44th UN session, held in New York in April 2011, by the vice president of The National Population and Family Planning Commission of China (UN, Wang, 2011).

By comparison, the fertility rate of the 27 European Union countries was 1.6 in 2009 (Older, 2011).

The Human Development Index, developed by the United Nations, assesses life expectancy, income and educational component. Despite China currently being the second largest economy in the world, the counterpoint in terms of Human Development Index does not follow this development, as, in 2009, China was ranked in 92nd position among the assessed countries, corresponding to an average level of development.

Brazil, meanwhile, was ranked in 75th position, corresponding to countries with a high level of development, but without harmony regarding the place it occupies in terms of its economy in the global context.

There is a fundamental question to be asked: why do big economies, nowadays, not allow the citizens of the country to enjoy the benefits of its economic growth, with simultaneous social development for the population?

Sustainability, Globalization and Development

Sustainable production is defined by Kriebel, Geiser and Crumbley (2001, p.298) as "the creation of goods and services using processes and clean systems, that conserve energy and natural resources, economically efficient, safe and healthy for employees, communities and consumers, and, socially rewarding and creative for all workers".

The effects of globalization have been and are devastating, with increasing concentration of wealth in countries with different levels of development.

Monteiro, Siqueira and Corrêa Filho (2008) report that "in macro perspective reported by Partanen, Loría-Bolaños, Wesseling et al. (2005, p.313) the current scenario presents a 'declining trend in the autonomy of national governments, minimization of the state and social well-being, reduction of social security, increased intensity of work [...]', among others. It impacts on the development of public policies related to labor, education, health and environment, which are fundamental for the development of the country".

In this context, the 'Principle of precaution in decision making', proposed by Joel Tickner, in 1998, in the *paper* 'A common sense framework for operationalizing the precautionary principle' should be noted:

"– Responsibility: when starting a new activity, the person who starts the process is responsible for proving that there was no safer alternative to accomplish it;
– Respect: in conditions of serious risk, preventive action is needed to prevent damage, even if there is no scientific evidence for the cause-effect relationship;
– Prevention: there is a duty to develop means to prevent any potential damage, rather than seeking to control and manage them later;
– Obligation to know and to inform: there is a duty to understand, investigate, inform (especially those who are potentially at risk), and act on potential impacts, that cannot be put aside by ignorance;
– Obligation to share power: democratization of decision-making in relation to science and technology" (Riechmann, 2002, p.25).

Sustainability, in addition to the slogan used in marketing by governments and companies, this should be seen as a challenge in the present, to preserve natural resources and ensure its proper use, for current and future generations. Corporate social responsibility should, in fact, have the sustainable production process, fair and honest marketing to population and care for the environment in which the company operates as a reference, among others.

Another aspect to be considered, from a philosophical perspective, besides the material aspect, is that work is fundamental to people's lives and enables them to develop their skills and have a relationship with the world, in addition to obtaining a livelihood for them and their family. From this perspective, it is important to highlight the point made by Wooding and Levenstein (1999, p.143), namely that "[...] we shape the world. Work is our creativity, our imagination, the way we demonstrate our commitment to the community and to its development. Work feeds our bodies and spirits".

Discussions on well-being at work have been widespread throughout Europe, at the end of the first decade of the 21st century, as have the benefits of its implementation that extend to the national economy and to the level of companies and individuals. In terms of the national level, these benefits can be characterized by increased productivity, more time at work and later retirement. Regarding the company, the highlighted items are the company's image, continuing education, profit, quality, competitiveness, mutual respect and initiative. From the individual perspective, it can contribute to the improvement of career, motivation, more appropriate work and distribution of free time, besides decreased stress levels (Antonen; Räsänen, 2008).

The United Nations Environment Programme states that the "creation of a sustainable economy will require a new political framework, based on equity and social solidarity [...] A sustainable economy must be *green for all*, creating decent jobs and stable communities and enabling fair distribution of wealth" (United Nations..., 2008a).

Social dialogue between workers, employers and government, aiming at fair policies that counterbalance costs and benefits are important aspects to be observed in the transition to sustainability (United Nations..., 2008b).

Kriebel, Geiser and Crumbley (2001, p.298) support the thesis that "many productive processes can be clean, safe and profitable, if they are delineated from the beginning, incorporating these criteria".

These authors also defend nine principles for implementing programs in sustainable production:

1. "Products and packaging are designed to be safe and ecologically sound throughout their life cycle, and services are designed to be equally safe and sound.

2. Waste and ecologically incompatible by-products are reduced, eliminated, or recycled.
3. Energy and materials are conserved, and the forms of energy and materials used are most appropriate for the desired ends.
4. Chemical substances, physical agents, technologies, and work practices that present significant hazards to human health or the environment are eliminated" (Kriebel; Geiser; Crumbley, 2001, p.299).
5. "Workplaces are designed to minimize or eliminate physical, chemical, biological and ergonomic hazards.
6. Workplace decision-making is an open and participatory process of continuous evaluation and improvement, encouraging the long-term rather than short-term economic viability of the enterprise.
7. Work is organized to conserve the efficiency and enhance the creativity of all employees.
8. The security and well-being of all employees is a priority, as is the continuous development of their talents and capacities.
9. The communities that surround workplaces are respected and enhanced economically, socially, culturally, and physically with a commitment to social equity and fairness" (Kriebel; Geiser; Crumbley, 2001, p.299).

The text written by Kriebel, Geiser and Crumbley (2001), published a decade ago, is still contemporary, both regarding the relevance of the theme discussed and the proposed guidelines, and even in terms of the slow incorporation of these principles in the design of production processes, in general.

From this perspective, the development of a sustainable society will benefit not only the current society, but also that of future generations.

Acknowledgements

To the Chinese and Brazilians students and colleagues and people with whom I lived in China – in Beijing and in Shanghai, during the China Top Program 2010, held in July 2010, which enlarged the vision for the country, its culture and uniqueness.

To the Chinese and Brazilians students and colleagues and people with whom I lived in Brazil, during the Brazil Top Program 2010, held in August 2010, which helped me to understand the intricacies of other public state universities in São Paulo and the partners in the project (USP and UNESP).

To Gabriela Cristina Granço Amaral, with whom I met during the Brazil Top Program 2010, who accepted the challenge of developing a scientific initiation given by CNPq, under my guidance, considering China as a reference.

To teachers Lin Chau Ming and Renata Cardoso Magagnin, from UNESP – Botucatu and Bauru, respectively, who coordinated the China Top Program 2010, for safe guidance abroad, of a heterogeneous multidisciplinary group, representing different Brazilian States.

To Prof. Dr. Leandro Russovski Tessler – Coordinator of the Coordination of Institutional and International Relations from the University of Campinas – Unicamp, for supporting activities during the China Top and Brazil Top – 2010 Programmes.

To University of Campinas – Unicamp for their support.

To my family for their encouragement.

References

AMARAL, G. C. G. *Análise das reformas econômicas chinesas e suas implicações no setor trabalhista*: as empresas estatais e seus trabalhadores. Relatório final de iniciação cientifica – CNPq, 2011.

ANTONEN, H.; RÄSÄNEN, T. (eds.) *Well-Being at Work:* New Innovations and Good Practices. Helsinki: Finnish Institute of Occupational Health, 2008.

INSTITUTO Brasileiro de Geografia e Estatística – IBGE. *Sinopse do Censo Demográfico 2010*. Brasília: IBGE, 2011. Available at: http://www.ibge.gov.br/home/estatistica/populacao/censo2010/sinopse.pdf. Access in: August 1st, 2011.

INSTITUTO Brasileiro de Geografia e Estatística – IBGE. *Síntese de Indicadores Sociais 2010*. Available at: http://www.ibge.gov.br/home/presidencia/noticias/noticia_visualiza.php?id_noticia=1717&id_pagina=1. Access in: April 2nd, 2011.

KRIEBEL, D.; GEISER, K.; CRUMBLEY, C. The Lowell Center for Sustainable Production: integrating environment and health into regional economic development. In: FERRANT, R.; PYLE, J. L.; LAZONICK, W. et al. (eds.) *Approaches to sustainable development*. The public university in the regional economy. Lowell: University of Massachusetts, 2001. p.295-308.

MONTEIRO, M. I.; SIQUEIRA, C. E.; CORRÊA FILHO, H. R. Produção justa e sustentável, meio ambiente e qualidade de vida corporativa. In: VILARTA, R.; GUTIERREZ, L. G. (org.) *Qualidade de vida no ambiente corporativo*. Campinas: Ipes, 2008. p.43-9.

RIECHMANN, J. Un principio para reorientar las relaciones de la humanidad con la biosfera. In: RIECHMANN, J.; TICKNER, J. (coord.) *El principio de precaución en medio ambiente y salud pública*: de las definiciones a la práctica. Barcelona: Icaria Editorial, 2002. p.7-37.

SOCIAL Agenda. *Older but healthier*. 2011. 27:7-8.

SOCIAL Agenda. *Greening the global economy*. 2011. 27:25.

UNITED Nations Commission on Population and Development. April, 2011. Available at: http://www.un.org/esa/population/cpd/cpd2011/countrystatements/agendaitem4/china.pdf. Access in: August 1st, 2011.

UNITED Nations Development Programme. *Human development Report 2009*. Available at: http://hdr.undp.org/en/statistics/indices. Access in: July 1st, 2010.

UNITED Nations Environment Programme – UNEP. *Green jobs*: towards a decent work in a sustainable, low-carbon world. Washington: UNEP, ILO, IOE, ITUC, 2008a.

UNITED Nations Environment Programme – UNEP. *Green jobs*: towards a decent work in a sustainable, low-carbon world. Policy messages and main findings for decision makers. Washington: UNEP, ILO, IOE, ITUC, 2008b.

WOODING, J.; LEVENSTEIN, C. *The point of production*. Work environment in advanced industrial societies. New York: The Guilford Press, 1999.

6

Strengthening Regional Capacities for Providing Remote Sensing Decision Support in Brazil in the Framework of Top China

Humberto Alves Barbosa[1]

Introduction

Since 2009, the Top China Programme has been a forum in which to share experiences, expectations and new ideas for applying emerging strategies by students and professors from Brazil and China so that they can discuss topics of global interest. The program provides opportunities for learning from others who work both within and outside these disciplines, from professionals who all face similar challenges and share a similar passion for improving their practice. It has served to build international working relationships that continue to have multilateral research opportunities for international projects. In the 2011 edition, the Programme extended this tradition to Chinese host universities: Peking University and Shanghai Jiao Tong, where 100 students and 22 professors deepened their knowledge over three weeks.

In the 2011edition, each of the 22 Brazilian universities appointed 100 students and 22 professors to collaborate with the Top China Programme. Supported by the Santander Universities, the São Paulo State

---

1 Universidade Federal de Alagoas (UFAL), Laboratório de Análise e Processamento de Imagens de Satélites (LAPIS), Maceió, Alagoas, Brazil. E-mail: *barbosa33@gmail.com*.

University (Unesp) implemented and coordinated scientific activities that were organized according to three thematic issues: environment, sustainability and urbanism. The main objective of the technical assistance was to effectively coordinate and communicate between different constituents and to promote the visibility of the Programme.

The Federal University of Alagoas (UFAL), through the Laboratory for Analyzing and Processing Satellite Images (LAPIS), was responsible for promoting research on environmental applications during the 2011 edition. LAPIS is responsible for developing methodologies and tools to monitor and assess crop productivity and convective storms derived from different satellites in South America (Barbosa et al., 2012). The essential activities for LAPIS are: data collection, mapping, information exchange, organization of technical and scientific meetings and knowledge dissemination.

The focus of this chapter is therefore to strengthen the capacities of the regional scientific community and provide stakeholders from agricultural activities with satellite information that is directly useful to increase the capacity of universities to develop methodologies and tools to monitor and assess environmental conditions in the context of the Top China Programme. This chapter is based on a case study from Brazil that demonstrates the use of satellite imagery data, obtained from free or low cost platforms, to provide information to address the aforementioned environmental issues, which are particularly critical in developing countries. Based on the expertise, experience and interest of researchers from LAPIS, this chapter focuses on harnessing satellite remote sensing resources and products to assess and monitor agricultural productivity.

The following paragraphs include a case study where a time series of satellite imagery is used so as to provide a large impact on agrometeorological researching worldwide:

i. Sugarcane yield estimation and modelling based on Normalized Difference Vegetation Index (NDVI) data over Southeastern Brazil and ii) SEVIRI/Meteosat biophysical products provided by the Laboratório de Análise e Processamento de Imagens de Satélites (Laboratory of Analysis and Processing of Satellite Images, LAPIS) (http://www.lapismet.com) in Brazil.

The case study was adopted within the framework of 2011 Top China, which contributed to applying emerging technologies and strategies in agrometeorological researching and training.

Quantification of sugarcane crop productivity: A study case in Southeastern Brazil

Agriculture represents an important segment of the economy in Brazil. Over the past 30 years, Brazilian agricultural growth and development has been guided by policies and technologies based on research for development. Brazilian policies, experiences, and focuses on social programs, such as *FOME Zero* (Zero Hunger), target social protection networks for the poor, thereby providing an important link between the regions (FAO, 2003).

Overall, current management practices in the Brazilian field of agricultural management are still very often based on outdated knowledge and technology. Similar to what happens in many other regions of the world, scientists frequently plan and develop their methods in isolation, without grasping what was really required by the relevant stakeholders (Barbosa; Lakshmi Kumar, 2011). On the other hand, stakeholders are often unaware of what science or knowledge-based alternatives are available. Scientific research is further isolated by a lack of proven utility and inadequate representation of results, whilst agricultural policy and management is isolated by legal and professional precedence. In the scientific community, model performance is frequently evaluated using procedures and statistical indicators which do not necessarily reflect usefulness for practical decision-making in the region of interest. Improved awareness of stakeholders' real needs can help scientists better orient their work. On the other hand, the typical current performance rates of crop modeling applications in the region may not be sufficient for practical decision-making. Poor performance is often a consequence of limited input data sets on which the model application is to be based. However, over the past decades, an increasing amount of potentially useful products derived from remote sensing have become available, but their potential for improving model performance in the region has to be well assessed (Barbosa; Lakshmi Kumar, 2011). Remote sensed imagery plays an important role in agricultural crop production over large area, both quantitatively and non-destructively, because agricultural crops are often difficult to access, and the cost of estimating ground productivity can be high.

The recent development of GEONETCast-EUMETCast data has allowed us to obtain frequent and accurate measurements of a number of basic agrometeorological parameters (e.g. evapotranspiration, surface albedo, surface temperature, solar radiation, rainfall, etc.). The GEONETCast-EUMETCast real-time and on-line data dissemination systems represent a global network of satellite-based data dissemination systems designed to distribute space-based, air-bone and *in situ* data, metadata and products to diverse communities. The satellite estimated agro-meteorological parameters (Barbosa; Huete; Baethgen, 2006; Barbosa; Lakshmi Kumar, 2011) provide a complete and spatially dense observation for assessing regional vegetation potential productivity. With a Geographical Information System (GIS) approach, it provides a framework to ingest the information of agro-meteorological parameters that influence crop yield.

To determine agriculture productivity, this case study aimed to develop a GEONETCast-EUMETCast product-based method of estimating the productivity of sugarcane using an agrometeorological spectral model. This study was carried out in the Municipalities of Barretos and Morro Agudo, located in the state of São Paulo, Southeastern Brazil (Figure 1). The analysis was performed for the crop from 2009/2010 and 2010/2011.

The values from the sugarcane parameters used, such as Respiration Factor (RF) (0.5 for temp. ≥ 20°C and 0.6 for temp <20°C); Agricultural Productivity Factor (APF) (2.9), Yield Response Factor (Ky) and Crop Co-efficient (Kc), were taken from Doorenbos and Kassam (1979) and Gouvea (2008). The EUMETCast service is installed at the Laboratory of Analysis and Processing of Satellite Images (LAPIS) at the Federal University of Alagoas (UFAL). The NDVI S10 remote sensing data, Production of Dry Matter (DMP) are available at the website http://www.lapismet.com and the SPOT Vegetation indices of VITO were collected from http://free.vgt.vito.be/.

This application proposes to test a remote sensing approach to quantify estimates of sugarcane productivity in Coruripe with the Integrated Land and Water Information System, (ILWIS, 3.7.1) GIS software. ILWIS was used to compute sugarcane crop estimates for each pixel in NDVI DMP and ETp images by applying radiative, aerodynamic and energy balance physics in 7 computational steps. These images are currently provided over both daily and 10-day composites at approximately a 3km and 1km spatial resolution, by EUMETSAT and VITO respectively.

Figure 1. Spatial variability of crop yields (2010/2011) for Barretos and Morro Agudo in Sao Paulo, Brazil.

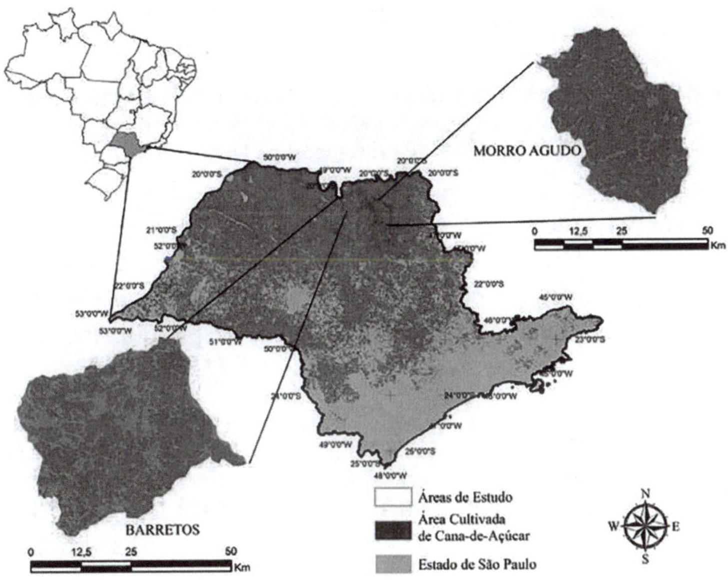

Step 1: Input NDVI and DMP databases using algorithm adapted from GEONETCast Toolbox

To implement a methodology for the ingestion of both the NDVI and DMP databases (raster) into ILWIS, specific routines of GEONETCast-toobox are adapted to import the datasets. For the ingestion procedure, based on a GIS approach using open source components, it requires additional work on corrections using overlays (Status map and LOG-file) to mask all appropriate areas of sugarcane crops in Coruripe.

Step 2: Computation of Fractional Vegetation Cover (FVC) from NDVI

For each pixel, the NDVI is converted into Fractional Vegetation Cover (FVC) by means of the formula set out by Munoz et al. (2005). The FVC is the one biophysical parameter that can determine the contribution partitioning between bare soil and vegetation for surface evapotranspiration, photosynthesis, albedo, and other flows crucial to land-atmosphere interactions.

$$FVC = 1.1101 * NDVI - 0.0857$$

Step 3: Computation of Leaf Area Index (LAI) from FVC

For each pixel, the FVC is converted to Leaf Area Index (LAI) by means of the formula from Norman et al. (2003). The LAI, defined as the total one-sided leaf area per unit ground area, is one of the most important parameters that characterize a canopy. Because LAI most directly quantifies the plant canopy structure, it is highly related to a variety of canopy processes, such as evapotranspiration, interception, photosynthesis and respiration.

$$LAI = -2Ln\,(1 - FVC)$$

Step 4: Computation of the growth factor from the LAI

Berka, Rudorff e Shimabukuro (2003) developed a simple approach for deriving the growth rate equation from the LAI. Experimental evidence indicated that the growth rate of several agricultural crop species linearly increases along with increasing amounts of LAI, when soil water nutrients are not limited (Doorenbos; Kassam, 1979). The following equation is used:

$$CGF = 0.515 - e[^{-0.667 - (0.515 * LAI)}]$$

where CGF = Corrected Growth Factor. Experimental evidence indicated that the growth rate of several agricultural crop species linearly increases along with increasing amounts of LAI, when soil water nutrients are not limited.

Step 5: Computation of the maximum yield potential (Yp)

The final equation that was used to derive maximum yield potential (Yp) includes evaporative fraction corrected growth factor (CGF), respiration factor (BF), agricultural productivity factor (APF) and dry matter production (DMP).

$$Yp=CGF*BF*APF*DMP$$

Where Yp is the maximum yield potential (kg ha^{-1}).

Step 6: Retrieval of evapotranspiration (ETp) via Land Surface Analysis –Satellite Application Facility (LSA SAF) ETp product

The crop coefficient is defined as the ratio of crop evapotranspiration, ET_r, to reference evapotranspiration, ET_p. K_c is crop specific and ranges from zero to over unity, depending on the crop growth stage. Crop evapotranspiration, at any time during the growing season, is the product of reference evapotranspiration and the crop coefficient.

$$ET_r = ET_p * K_c$$

Crop coefficients was developed for nearly all crops by measuring crop water use with lysimeters and dividing the crop water use by the reference evapotranspiration for each day during the 2009/2010 growing season.

The recent development of LSA-SAF products has allowed us to obtain frequent and

accurate measurements of a number of basic agrometeorological parameters (e.g. surface albedo, surface temperature, evapotranspiration). The satellite estimated agrometeorological parameters have several advantages compared to conventional measurements of agrometeorological data in the ground meteorological network.

Step 7: Estimation of sugarcane productivity

The sugarcane yield estimation model over the growing season, on a biweekly basis, is accomplished by using an agrometeorological model integrated to the ILWIS, in accordance with Doorenbos and Kassam (1979):

$$Y_e=Y_p[1 - ky(1- ET_r/ET_p)]$$

where Ye is the estimated yield (kg ha^{-1}), Yp the maximum yield (kg ha -1), ky the yield response factor; ETr the actual evapotranspiration (mm) and ETp the maximum evapotranspiration (mm). Maximum yield (Yp) is established by the genetic characteristics of the crop and the degree of crop adaptation to the environment. The resulting map of the estimated yield (Ye) is clipped to mask the Barretos and Morro Agudo municipality boundaries in the State of São Paulo, Brazil. In order to establish correct coordinates, a map calculation within the ILWIS is used to implement this procedure. The flow diagram of the methodology to quantify sugarcane productivity via satellite products is shown in Figure 2.

Figure 2. Flow diagram of methodology of quantifying sugarcane productivity via satellite products.

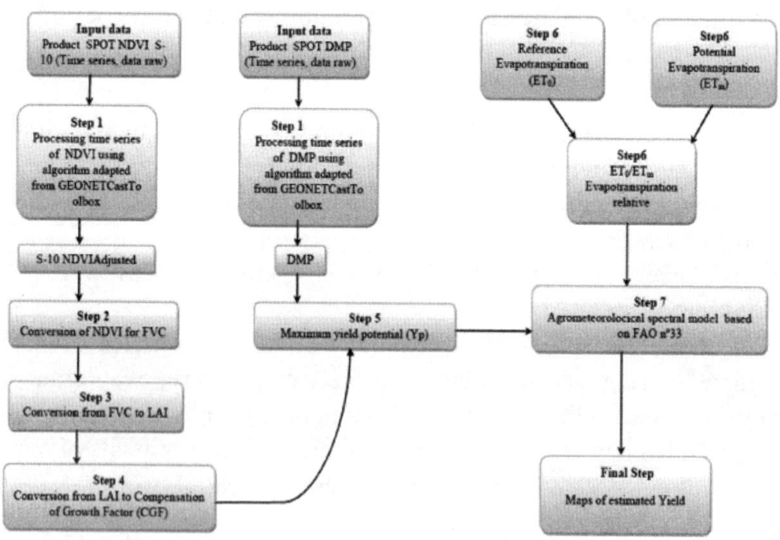

Figure 3 shows the spatial variations in sugarcane production in Barretos and Morro Agudo for 2009/2010 and 2010/2011. The figure clearly indicates high spatial patterns in yield variability. This could be due to the mixing of a significant fraction of the observed pixels for the "arable pixel" and "non-arable pixel" within the municipalities. The quantified results give sugarcane yield mean range of 50 to 135 Ton ha^{-1}. The results obtained here represent a first step towards the operational use of ILWIS tools in Brazil using NDVI S-10, DMP SPOT and ETo for operationally estimating sugarcane productivity. Overall, the model was able to identify and quantify (Table 1) the spatial variability of agricultural production in the municipalities analyzed. Therefore, the methodology is useful for developing estimates of operational support for sugarcane productivity (Barbosa et al., 2012).

Figure 3. Spatial variability of crop yields in Barretos and Morro Agudo for 2009/2010 and 2010/2011.

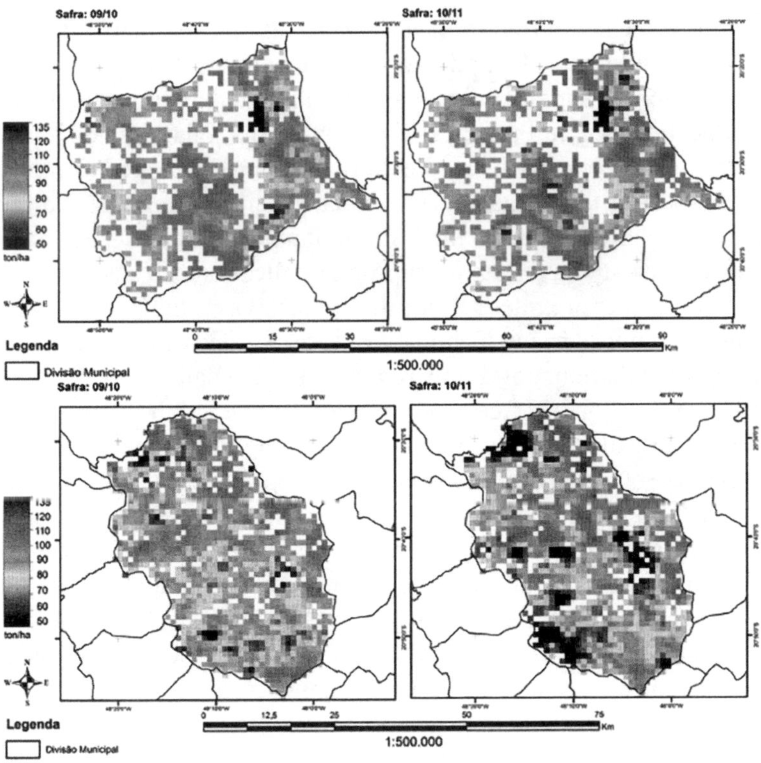

Table 1. Comparison between the productivity of sugarcane using an agro-meteorological spectral model and harvested crop yield from National Food Supply Company (CONAB).

Mean Productivity (t/ha)		
Barretos	Crop	
	2009/2010	2010/2011
Agrometeorological Model	93,15	93,31
CONAB	96,21	93,29
Mean Productivity (t/ha)		
Morro Agudo	Crop	
	2009/2010	2010/2011
Agrometeorological Model	93,15	93,31
CONAB	96,21	93,29

Outlook and conclusions

The Top China Programme has served as an exemplary endeavor and is providing a role model to increase and improve Brazil-China cooperation in support of transferring knowledge and experiences in education and best practices through international partnerships. A scientific network for this cooperation is being developed to capture the support of the Santander Universities for building international working relationships. All the information produced has made available to interested universities through publications.

This chapter provides a multi-disciplinary case study on the power of using remote sensing technologies to assess and monitor crop productivity. It demonstrates how remotely sensed data that is available in the public domain freely or at very low cost can be harnessed to address critical challenges in Brazil pertaining to agrometeorological research. Another valuable impact is the importance of joint cooperation with international research partnerships in order to encourage scientific exchange. However, despite the availability of the technology, full benefits from available remotely sensed imagery resources for developing countries can only be realized when enabling policies are formulated and implemented, and concerted capacity development is undertaken to establish a critical human resource base. This will enable the policy makers to go for the risk managing practices such as agricultural crop reinsurance schemes.

Acknowledgements

The author wishes to thank the Laboratório de Análise e Processamento de Imagens de Satélites (Laboratory of Analysis and Processing of Satellite Images, LAPIS) (http://www.lapismet.com) EUMETSAT for making the SEVIRI/Meteosat biophysical products publicly available. The professor and students participants of UFAL in the 2011 Top China framework are most grateful to the Santander Universities for funding this programme.

References

BARBOSA, H.; HUETE, A.; BAETHGEN, W. A 20-year study of NDVI variability over the Northeast region of Brazil. *Journal of Arid Environ*, v.67, p.288-307, 2006.

BARBOSA, H. A.; LAKSHMI KUMAR, T. V. Strengthening regional capacities for providing remote sensing decision support in drylands in the context of climate variability & change. In: YOUNG, S. S.; SILVERN, S. E. *Environmental change*. USA, 2011. ISBN 979-953-307-109-0.

BARBOSA, H. A. et al. Earth Observation and satellite information for monitoring desertification, drought and agricultural activities in developing countries? In: YOUNG, S. S.; SILVERN, S. E. Environmental change. USA: Salem State, 2012. ISBN 980-953-307-806-4.

BERKA, L. M. S.; RUDORFF, B. F. T.; SHIMABUKURO, Y. E. Soybean yield estimation by an agrometeorological model in a GIS. *Scientia Agricola*, v.60, n.3, p.433-40, 2003.

DOORENBOS, J.; KASSAM, A. H. *Yield response to water*. Roma: Food and Agriculture Organization of the United Nations, 1979. 193.

EUMETSAT. A planned change to the MSG Level 1.5 image product radiance definition, document n.519. 2007a. Available at: *http://www.eumetsat.int/idcplg?IdcService=GET_FILE&dDocName=PDF_MSG_PLANNED_CHANGE_LEVEL15&RevisionSelectionMethod=LatestReleased.*

GOUVEA, J. R. F. 2008. *Mudanças climáticas e a expextiva de seus impactos da cultura de cana-de-açúcar na região Piracicaba-SP*. Dissertação (Mestrado em Física do Ambiente Agrícola) – Escola Superior de Agricultura "Luiz de Queiroz", Universidade de São Paulo, Piracicaba 99.

MUNOZ, J. C. J. et al. *Fractional Vegetation Cover Estimation from Proba/CHRIS Data*: Methods, Analysis of Angular Effects and Application to the Land Surface Emissivity Retrieval. 3rd ESA-CHRIS/PROBA Workshop, ESRIN, ESA SP-593, Italy, Frascati, 2005.

NORMAN, J. M. et al. Remote Sensing of Evapotranspiration for Precision Farming Applications. In: IEEE International Geoscience and Remote Sensing Symposium, 2003, Tolouse. Proceedings: IGARSS 2003 – Organizing Committee, p.21-5.

RUDORFF, B. F. T. Dados Landsat na estimativa da produtividade agrícola da cana-de- açúcar. Dissertação (Mestrado em Sensoriamento Remoto) – Instituto Nacional de Pesquisas Espaciais – INPE, São José dos Campos, 1985.

7

One mode of Urban Agriculture: Organic Vegetable Production and Marketing

Huang Danfeng; Wang Xiaoli[1]

Development of urban agriculture

Requirements of urban vegetable supply

The Chinese urban population is about 600 million at present, which accounts for 45% of the national population. Expert forecasts state that the Chinese urban population will reach 926 million by 2025. Eight metropolitan areas, Chengdu, Chongqing, Guangzhou, Shenzhen, Tianjin and Wuhan, as well as Beijing and Shanghai, will populated with more than 10 million people, an increase of nearly four times, which accounts for 13% of the national population. These metropolitan areas will reach world-class city status on the same line as London, New York and Tokyo (*CHINESE consumer survey report 2012*). In the next 20 years, these eight Chinese metropolises will bring heavy demands for food, especially in terms of the supply of grain and vegetable (Lei, 2012).

Urban agriculture is defined, by Food and Agriculture Organization (FAO) and Agency for Planning Development in 2003, as an industry that responds to the nutritional demands of a city. Its main purpose is

1 School of Agriculture and Biology, Shanghai Jiaotong University, Shanghai 200240.

to satisfy the needs of urban consumers. The supply of fresh agricultural products is the major problem of maintaining urban operating safety and is also presently the leading function of urban agriculture in China. Under the process of urbanization, urban residents have more and greater demands in terms of the quantity, variety and quality of agricultural products (Lei, 2012).

The supply of vegetables is the main composition of the supply of subsidiary foods in urban areas. It has played an important role in stabilizing commodity prices and securing people's lives to ensure the supply and security of vegetables. It is expected that the number of wealthy households in China will be over 4.4 million in 2015, next only to the USA, Japan, and England. The bulk of this population is concentrated in the metropolis (*CHINESE consumer survey report 2012*), their increased purchasing power will promote their high demands for quality and security of vegetables. Therefore, the theme of urban agriculture development is to excavate the function of production, ecology and life in agricultural production and find a suitable agricultural production pattern for urban development.

Development of organic vegetable production

Organic agriculture, developed in the 1970s, is a kind of agricultural production. It conforms to the modern health concept, drawing on the experience of traditional agriculture and is applied to a series of modern agricultural technology. The International Federation of Organic Agriculture Movements (IFOAM) established four principles for the development of organic agriculture: health, ecology, justice, and care. The State Environmental Protection Administration of Organic Food Development Center (OFDC) proposes a definition of organic farming, namely that organic farming is a kind of agricultural production, following the natural rules and ecological principle, balancing the coordination of planting and breeding by using a series of the sustainable development of agricultural technology and maintaining the stability of agro-ecosystem (Liping, 2012b; Lehne; Shiwei; Danfeng, 2006; The Principles of Organic Agriculture).

The area of global organic farming was up to 37 million hectares by the end of 2010, mainly distributed across Oceania, Europe and

South America, accounting for 38.7%, 22.6% and 19.0% of the world's total organic agricultural area, respectively. The three largest countries that have organic agricultural production areas are Australia (12 million hectares), Argentina (4.18 million hectares) and the United States (1.95 million hectares). In China, 1.39 million hectares are under organic management, while the productive area for organic vegetables accounts for about 11,600 hectares (The Principles of Organic Agriculture; Soel, FiBL, 2012).

Recently, organic food is popular among all consumers because of its safety. The total sales of global organic food is US$59.1 billion in 2010, which is three times more than in 2000. Consumption of organic food in North America and Europe accounts for 96% of the world's total.

In China, the production scale and market share of organic food is very low, still generally in its initial stage. Certified organic products include grain, vegetables, fruits, tea, bee products, natural spices, herbs, dairy products, livestock products, wild products, aquatic products and so on. The sale of organic vegetables are slightly less than grain. In 2009, the domestic sale of organic food reached US$1.587 billion. The export volume is 0.46 billion and only accounts for 0.7% of the international market (Soel, FiBL, 2012).

Challenges and Opportunities

The economic and urban development in China provided the Hugh development with the opportunity for organic farming, meanwhile, the process management for organic farming production and the credit system in marketing organic produce are challenging. The organic certification and production process control actually provide basic protection for the quality and safety of organic vegetables. However, many researchers, producers and consumers who are concerned with product quality, ecological security and sustainable agriculture still have questions. Is vegetable quality truly being improved when no use manufactured (synthetic) fertilizers, pesticides and transgenic seed are used? Is vegetable safety being more effectively guaranteed when organic farming use organic fertilizer and biological pesticide instead of chemical fertilizer and chemical pesticide? Is the production environment such as soil, water resource being under effective protection? Does the price of organic vegetables fully reflect the quality and safety of products? All these questions require more studies and time to be verified.

Consideration on the development of urban organic vegetables

The quality and safety of organic vegetables

Organic production systems come into being. Conventional agriculture or chemical agriculture can make use of cheap fossil energy, industrial technology or equipment to improve labor productivity and increase crop yield. It leads to energy losses, natural resource depletion, environmental pollution, ecological damage, and other serious problems (Liping, 2012b; Lehne; Shiwei; Danfeng, 2006). Furthermore, long-term conventional management can cause many environment problems, such as nitrogen leaching, groundwater pollution, destruction of soil structure and pesticide accumulation (Ge et al., 2010; Fenghua; Wenhua; Jianling, 2009; Ge et al., 2006; Ge et al. 2008). As regards the food security problem, for example, excessive pesticide residues, high nitrate levels in vegetables and fruits/ food poisoning have attracted increasing attention to food safety. Thus, the central aim of agricultural production process management is to guarantee food security and protect consumer health. Organic agriculture is not allowed to use synthetic chemical substances such as pesticides, chemical fertilizers, growth regulators and feed additives, nor are genetically engineered organisms and their products, however, it follows the laws of nature and ecological principles. It adopts a series of sustainable agro technique, coordinates the balance between planting and breeding industry and thus can maintain the stability of agro-ecosystems (Song et al., 2010; Ge et al., 2011). Organic production can provide high-quality healthy food with no negative effects to the environment. It not only provides a sufficient number of high-quality foods, but coexists in harmony with nature. So the concept of organic agriculture production with technological systems comes into being.

Strengthening the control of the whole life cycle of organic production. Quality assurance of organic vegetables comes from the ecological management of the entire production cycle. The foundation of supply chain quality security systems in organic vegetables is the basis for obtaining high-quality organic vegetables and safeguarding quality and safety. The ecological conditions for the production base are the basis of producing organic vegetable. Pure water can effectively activate nutrients, making plants absorb enough water and nutrients; soil physical and chemical properties, nutrient balance conditions, determine the organic production fertilization strategy and soil-borne diseases biological prevention or feasibility. Strengthening the management of the production process from seed to harvest is the key to protecting the quality of organic vegetables. Healthy soil and healthy plants can ensure product quality. Postharvest preservation of physical processing, shortening the logistics time etc, makes it possible to maintain the quality of fresh vegetables when they are on the consumer's table.

Organic vegetables does not equal quality and safety. An organic management system is an ecological production model, which emphasizes low-carbon management, biodiversity, ecological balance, especially regarding management during the production process rather than the detection of end products, so that the organic food is more trustworthy in terms of security. However, there are doubts: do organic vegetables really have a higher quality than conventional ones, or are organic vegetables safer than conventional ones? On the one hand, some quality management system such as GAP, EUREP-GAP and QACCP can fully ensure the quality and safety of vegetables (Danfeng, 2012); on the other hand, some kind if scientific management in the organic system needs to be developed in practice, such as the transformation and balance of soil nutrients, maintenance of microbial diversity, protection against pests, and so on.

Organic production and ecological security

Organic production is a system. Organic production runs through the entire production and supply process from the beginning of a suitable production environment selection; it is a system which makes the agricultural production and ecological cycle coexist harmoniously from animal, plant and microbe. Compared with conventional production, the biggest change in the organic vegetable production system is the management theory. Their focus is not only the crop yield and crop quality, but also the quality and safety of the agro-ecosystem.

Low carbon concept is practiced in organic farming. During organic production we can improve the plants' resistance to diseases and insects themselves by selecting and breeding resistant varieties rather than using chemical pesticide. Many studies indicate that the soil's microorganism plays an important role in organic N transformation, nitrate nitrogen supply and maintaining the balance of the ratio of carbon and nitrogen in organic plant production. It is well known that a plant can directly absorb organic N, but whether plants can efficiently absorb and assimilate organic N in an organic system is still unclear (Shiwei, 2009a; Shiwei, 2009b; Ge et al., 2009). The use of low-carbon organic matter may increase the soil potassium use efficiency and enhance soil microbe activities (Ruijuan et al., 2009). Thus, improving plants' resistance, enhancing soil microbial diversity and offering a low-carbon organic production environment is at the center of organic agro-ecosystem management theory.

Ecological effects of organic production are mentioned. The major aim of the rejection of genetically engineered plants and animals is to maintain the natural ecological balance and prevent their disturbance in nature. Avoiding the use of chemical pesticide can protect the biodiversity, decrease the drug resistance of animal and plant, and reduce pesticide residue in the soil and water. Therefore, the organic production process is more consistent with the nature principle and ecological theory; it is a sustainable agricultural production type that can maintain the persistent stabilization of agro-ecosystems.

Organic production is a production type that can supply high quality vegetables, without affecting the environment and living in harmony with nature (Lehne et al., 2008). However, the effect of organic vegetable production on the environment requires a long-term follow-up investigation. Many questions still need to be answered, such as, nutrient balance of organic fertilizer input, changes in the soil microbial diversity and interaction between plant, animals and microbe.

Quality and price of the organic vegetable

Market researchers have shown that the price of organic vegetable is 2-5 times higher that that of conventional vegetables. Compared to tea, coffee, milk and grain, the price of organic vegetables is more reasonable. The production base of organic vegetables needs a long time to construct, including 2-3 years in the organic transitional stage. Monitoring and improvement of the soil and water source will increase production costs. The production process management, such as the production and utilize of organic fertilizer, pest, disease and weed control, also greatly increases production costs. The costs of logistics and marketing in the VIP distribution service from field to table should also not be neglected. Consumers are more concerned about product quality and company reputation than only price.

The high input in the process of organic transition, production and the cost of site inspection and certification lead to the price of organic vegetables increasing. The high cost also causes many small – and medium-sized enterprises to exit market. On the one hand, organic farming is of great benefit in the long term, its ecological and social benefits are greater than its economic benefits; On the other hand, counterfeit organic goods damage the market and make consumers feel suspicious about organic products. Is there a method that can discern the false from the genuine, or can reduce the certification cost by setting up a produce alliance with which one production enterprise only produces one single or a small number of products?

Consideration on urban organic vegetable development

Urban organic vegetable production

The development of urban organic vegetable production is limited by many factors, such as labor, land and resources. Due to the current situation of the spiraling costs of manufacturing materials, the increasing labor wages and the decreasing agricultural contrast benefits, the development of urban organic vegetable production must rely on scientific and technological innovation, and explore a new development model.

By using the e-commerce platform and the technology from the "EPC System Network," integrating the production bases of organic production, optimizing the layout of production and thus forming a idealized situation, which can be called "Each Enterprise Got a Special," can effectively reduce certification, technology and logistics costs; combining breed and cultivation in order to establish a low-carbon production system that can maintain the stabilization of organic agricultural ecosystem.

With the help of a highly efficient production mode, such as the mechanization and automation of organic substrate cultivation, enterprises with high funding can decrease labor input, intensity and improve production efficiency; establishing external bases and arranging production according to the characteristics of natural conditions will extend the chain of production and supply for urban organic vegetables.

Key technology of ecological management

Organic production areas are certificated. The first choice for the cultivate base of organic vegetables is the ecological agriculture cultivating area that is far from the pollution source. For instance, the construction of cultivate base on Chongming ecological island is consistent with the local ecological agricultural development, and can maintain the stability of the base construction.

Based on organic certification, recording the archives of the production environment, evaluating the environment in real-time or at a fixed period to make sure the requirements of organic production are met. By applying the "EPC System Network," the large organic agricultural enterprises can monitor and manage their production base during the whole production process.

Ecological management standards are established. In March 2012, the national standard (GB\T19630-2011, 2011; Liping, 2012a) was updated and implemented by the Certification and Accreditation Administration of the People's Republic of China. The quality standard of environment cultivation and the safety standard of input products can be assured by certification. By using instructions for technical operation, the key links, such as, variety selection, crop rotation, fertilization, irrigation, P&D control, postharvest and cold chain will be strictly and precisely completed according to the requirements of the organic production management. Establishing a field archive to ensure the product traceability. In order to successfully implement the national standard of organic products, the growth environment and internal cycle between the vegetables and animals needed to be empathized, thereby ensuring their living space and natural requirement in a reasonable range, not only for safe food.

Efficient production technology is needed. Many studies reported that the crop yield of organic farming was lower than that of conventional (Liping, 2012b; Song et al., 2010). The major reason for the drop in yield is the low nitrogen supply (Ge et al., 2006; Ge et al., 2009; Ruijuan et al., 2009). When supplied with sufficient organic fertilizer, the yields of strawberry, lettuce and onions from organic production were 11%~84% higher than those from conventional. The increase in soil microbe quantity and the improvement in soil properties (Ge et al., 2011; Ruijuan et al., 2009) under organic farming may explain this. Thus, there is no close relationship between crop yield and the cultivation system. By investigating the principles of nutrient cycle and constructing a more balanced fertilization practice, organic farming can also grain the same yield as conventional farming. In organic protected horticulture, with the help of the high-tech, automatic production system, e.g. organic ecological pattern, we also can gain high production efficiency.

Ecological management is used in all procedures. The concept and technology of ecological management should be run through the entire process of organic vegetable production. Ecological management consists of three parts: variety selection, soil management and fertilization, and pest, disease and weed control. Firstly, selecting high resistant, non-transgenic varieties with no chemicals treatment. Secondly, implementing scientific soil management and rational fertilization, limiting the fertilizer input to reduce nutrient loss and avoid water eutrophication. Making the fertilizer source clear to minimize the risk of heavy metal contamination in the product. Thirdly, by using natural predators, variety selection, crop rotation and temperature control and so on, we can make the growth conditions unsuitable for pests, disease and weeds. Physical control measures, such as using light and color to kill pests and artificial / mechanical weeding are encouraged in organic farming.

Certification and supervision are important. In order to strengthen the supervision of organic foods, the Certification and Accreditation Administration of the People's Republic of China (CNCA) announced "The rules of organic product certification and implementation" on December 2nd 2011 (GB\T19630-2011, 2011). Since July 1st, 2012. The new rules formulated that the minimum sales packaging of organic foods or products must be labeled with the unified symbol of organic food certification mark, which has a unique serial number and certification authority. The unique serial number consists of certification authority codes, time codes and random codes of authentication marks. Every organic product reaching the new national standard should be labeled with 17 unique serial numbers, "one product one number," no secondary packaging; the product quality can be traced back the whole process.

Marketing Modes of urban organic vegetable

There are great differences between the marketing modes of urban organic vegetable and the general vegetable supply. Membership distribution and group buying are the two most common ways of marketing, while direct monopoly marketing, organic restaurants, agricultural organization and residence docking marketing are gradually developing. E-commerce also provides a platform for urban organic vegetable marketing.

Membership distribution should be an effective way to meet the needs of consumer groups for organic vegetables, because they have the following characteristics of high-incomes, critical requirements of food quality and safety, environmental awareness and great working pressure. Membership distribution requires large continuous supplies of varieties of fresh and high-quality organic vegetables for the organic companies. At the same time, organic companies should have a good reputation, a well-developed logistics system and continuous supplying capabilities. Group buying of organic vegetables is a good choice for large enterprises to provide their employees with some benefits and the ancillary services orders in the high-end real estate purchase process, because they have a large number of continuous service requirements and general annual consumption. The orders for organic vegetables through membership distribution and group buying are relatively stable, which is beneficial for the organization and supply of vegetable production.

Direct Monopoly is a popular and mature marketing method around the world. Large supermarkets set organic vegetable sales over the counters or in stores around upscale communities, bustling commercial areas and business clubs. Organic restaurants serving consumer groups such as organic food advocate that white-collar workers and foreign guests also have a good market in these areas.

Urban residents who work and live in bustling metropolises have a very strong demand for organic farming and enjoying natural produce. Rural residence docking, sentinel production of organic vegetables, own labor during holidays, weekday hosted monitoring from a virtual farm to field planting has gradually become the way that come urban residents look to buy organic vegetables. Through agricultural tourism, some of the public have reached an agreement with the organic farm regarding vegetable production and the anniversary of supply, leasing farmland, agreeing production varieties and farming personally on holiday, leisure, day-to-day process by online video, the quality and safety are straightforwardly assured. In the international and domestic markets, the production model that pairs farmers and residents with rural and community interaction enriches the content of organic vegetable production and marketing.

With the rapid development of e-Business, online orders and third-party logistics are gradually becoming one of methods to market organic vegetables. The e-commerce platform of organic vegetables, which is guaranteed by a powerful logistics system, is greatly convenient for urban rural residence who wish to purchase.

References

CHINESE consumer survey report 2012 New consume strength. McKinsey Quarterly. Available at: *http://money.163.com/11/1221/11/7LQ01M5700253G87.html*.

DANFENG, H. The management principle and practice of QACCP for vegetable. *Journal of Changjiang Vegetable*, (8), p.1-5, 2012.

FENGHUA, Z.; WENHUA, L.; JIANLING, J. Effect of yield response and in taking nutritions of Chinese cabbage to excessive applying N, P and organic fertilizer. *Soil and Fertilizer Science in China*, (4), p.60-3, 2009.

GB\T19630-2011, Certification and Accreditation Administration of the People's Republic of China, China Zhijiang Publishing House, 2011.

GE, T. et al. Chemical properties, microbial biomass, and activity differ between soils of organic and conventional horticultural systems under greenhouse and open field management: a case study. *Journal of Soils and Sediments*, v.11(1), p.25-36, 2011.

GE, T. et al. Dynamics of Nitrogen Speciation in Horticultural Soils in Suburbs of Shanghai, *Pedosphere*, China, v.20(2), p.261-72, 2010.

GE, T. et al. Amino acids as a nitrogen source for tomato seedling: the use of dual labeled (13C, 15N) glycine to test for direct uptake by tomato seedling. *Environmental and Experimental Botany*, 66, p.357-61, 2009.

GE, T. et al. Effects of Nitrogen Forms on Carbon and Nitrogen Accumulation in Tomato Seedling. *Agricultural Sciences in China*, v.7(11), p.1308-17, 2008.

GE, T. et al. Research progress of the souble organic nitrogen in the production system of horticultural crop. Horticulture progress. *China Agriculture Press*, Beijing, (7), p.805-11, 2006.

LEHNE, P. et al. *Muskmelon growth, yield and nitrogen accumulation under organic and conventional farming conditions in Eastern China*. 16th IFOAM Organic World Congress, Modena, Italy, June 16-20, 2008. Archived at: *http://orgprints.org/view/projects/conference.html*.

LEHNE, P.; SHIWEI, S.; DANFENG, H. Ecological farming in Europe: a model for China? *Journal of Shanghai Jiaotong University* (Agricultural Science), 24 (supplement), p.33-40, 2006.

LEI, S. Practices of urban modern agricultural in Shanghai. Shanghai: Shanghai Scientific and technical Publisher, 2012.

LIPING, W. Explanation of new version "Implementation Rules for organic standards and Certification". *Journal of Changjiang Vegetable*, (11), p.7-12, 2012a.

LIPING, W. The background and Present Situation of organic farming. *Journal of Changjiang vegetables*, 9, p.7-10, 2012b.

RUIJUAN, S. et al. Variation of N & P Contents in Paddy Soil Water and Its Potential Environmental Effect under Pig Manure Application. *Soils*, (6), p.907-11, 2009.

SHIWEI, S. Difference in Nitrogen Uptake by Muskmelon in Organic and Conventional Farming Systems. *Chinese Journal of Applied and Environmental Biology*, 15(2), p.197-201, 2009b.

SHIWEI, S. Uptake Kinetics of Different Forms of Nitrogen by Muskmelon. *Chinese journal of Soil Scienc*, (3), p.590-3, 2009a.

SOEL, FiBL, Survey. Minou Yussefi, SOEL & Helga Willer, FiBL. *The World of Organic Agriculture*. Statistics and Emerging Trends, 2012.

SONG, S. et al. Yield, Fruit Quality and Nitrogen Uptake of Organically and Conventionally Grown Muskmelon with Different Inputs of Nitrogen, Phosphorus and Potassium. *Journal of Plant Nutrition*, 33, p.130-41, 2010.

THE PRINCIPLES of Organic Agriculture. Available at: *http://www ifoam.org/about_ifoam/principles/index.html*.

8

Plant Biotechnology: An Approach for Pesticide Reduction in Agriculture

Eliane Romanato Santarém[1]

Introduction

The term pesticide covers a wide range of compounds including insecticides, fungicides, herbicides, rodenticides, nematicides, plant growth regulators and others. Pesticides have been used successfully to control pests and plant diseases, resulting in an increase in agricultural output (Aktar; Sengupta; Chowdhury, 2009). However, the impact of pesticides on the environment is often greater than what was intended by those who used these agrochemicals. Over 95 percent of pesticides and herbicides reach a destination other than their target species, including non-target organisms ranging from beneficial soil microorganisms, to insects, plants, fish, and birds. Moreover, studies conducted by the U. S. Geological Survey (USGS) on major river basins across the country showed that more than 90 percent of water and fish samples from all streams contained one, or more often, several pesticides (U. S. Geological Survey, 1999).

---

1 Pontifícia Universidade Católica do Rio Grande do Sul, Faculdade de Biociências, Porto Alegre, RS, Brazil. E-mail: *esantarem@pucrs.br*.

There is now evidence that some of these chemicals pose a potential risk to humans and other life forms and cause unwanted side effects to the environment (Igbedioh, 1991). In fact, weed killers can be especially problematic because they are used in relatively large volumes. The impacts are mainly contamination of land and water, when pesticides escape from production sites and storage tanks, run off from cultivated fields, are improperly discarded, or are sprayed aerially. On the other hand, human contamination often happens by inhalation of aerosols, dust and vapor that contain pesticides, oral exposure by consuming food and water, as well as through dermal exposure caused by accidental contact of the pesticide with the skin. No segment of the population is completely protected against exposure to pesticides and the high risk groups exposed to pesticides include workers involved in pesticide formulation, sprayers, loaders as well as agricultural farm workers (Aktar; Sengupta; Chowdhury, 2009). Nevertheless, everyone may be exposed to some level of contamination when consuming water and food. Thus, it is vital that society finds a way to mitigate this problem. The balance of world food production and food requirements has to be achieved, and a sustainable agriculture is definitely required. The best way to reduce pesticide contamination and the harm caused by its use in our environment is to use safer, non-chemical pest control methods.

Biotechnology offers significant improvements in virtually every area of crop production and utilization, with potential benefits for farmers, the food industry, consumers and the environment. Biotechnology can contribute to the reduction of the use of chemicals in agriculture which represents fewer residues in the final product.

This chapter will mainly approach and discuss plant biotechnology, agriculture and environment protection through the reduction of the use of pesticides.

Biotechnology

The multidisciplinary character of biotechnology makes it rather difficult to define biotechnology. However, it could be defined as the application of scientific and engineering principles to the processing of materials by biological agents and to provide goods and services.

Historically, biotechnology is as old as human civilization and has been an integral part of human life. The term "traditional biotechnology" has been applied to the natural processes that have been used since ancient times to produce beer, wine, cheese and many other foods. Modern biotechnology refers to a number of techniques that involve the intentional manipulation of genes, cells and living tissue in a predictable and controlled manner to generate changes in the genetic make-up of an organism or to produce new tissue. Examples of these techniques include recombinant DNA techniques (genetic engineering), tissue culture and mutagenesis and all the improvements made in the traditional biotechnological processes.

The origins of biotechnology culminated with the birth of genetic engineering, with two key events that unite genetics and biotechnology. One was the 1953 discovery of the structure of DNA, by Watson and Crick, and the other was the 1973 discovery of a recombinant DNA technique by which a section of DNA was cut from the plasmid of a bacterium and transferred into the DNA of another. This approach could, in principle, enable bacteria to adopt the genes and produce proteins of other organisms, including human proteins. Popularly referred to as "genetic engineering," it came to be defined as the basis of new biotechnology.

Several products obtained by biotechnology have been developed and are already integrated in society. Transgenic plants with higher yield or that are resistant to herbicides and insects, as well as vaccines, diagnosis kits, genome sequencing of humans and several other species, transgenic animals for studying diseases, and environmental remediation are all examples of biotechnological products.

Biotechnology is expected to reach some achievements, such as modified biological organisms (plants, engineered yeasts and animals), production of therapeutic proteins and drugs aiming more efficiency at lower costs, modified human cells for treating genetic diseases, and environment protection or remediation.

Plant Biotechnology and Agriculture

Nowadays, biotechnology and genomics are the most promising tools for addressing new agricultural challenges. Agriculture of the future is based on four essential principles, which are as follows: the production of safe, high-quality food and feed in sufficient quantities, sustainable agriculture, the development of plants for the production of renewable resources and energy; increasing competitiveness while maintaining the freedom of choice for consumers.

Plant biotechnology involves techniques which allow the manipulation and modification of plant cells, tissues or organs. Plant tissue culture has now emerged as one of the major components of plant biotechnology. Considerable progress has been made to improve media formulations and techniques for culturing cells and tissues, resulting in *in vitro* regeneration of plants from several species, even those recalcitrant for tissue culture (Gupta; Ibarak, 2006). Recent development of *in vitro* techniques has demonstrated its application in the rapid clonal propagation, regeneration and multiplication of genetically manipulated superior clones, production of secondary metabolites and *ex-situ* conservation of valuable species. It has also been considered the basis of production for genetically modified plants, which have improved traits and characteristics.

Plant tissue culture

Plant cell culture, also known as micropropagation, is based on the unique property of the cell totipotency, which is the ability of a plant cell to regenerate into a whole plant. This property of the plant cells has been exploited to regenerate plant cells under laboratory conditions using artificial nutrient medium. This medium is prepared with mineral salts, vitamins and plant hormones, all of which are components needed by the plant to grow and develop.

Generally, micropropagation comprises four steps: induction of cell masses or shoots, multiplication (proliferation of shoots), rooting and elongation and acclimatizing (Figure 1).

Micropropagation starts by choosing the explant, which is the tissue obtained from a plant to be cultured, and may include portions of shoots, leaves, stems, flowers, roots and single, undifferentiated cells (Figure 1A). The explant is usually surface desinfested and cultured onto a specific medium for induction of shoots or cellular masses (Figure 1B). Once the shoots are obtained, proliferation should occur

(Figure 1B and 1C). An extra step of elongation for the shoots might be required for several species. Elongated shoots are subsequently induced to rooting and acclimatized in natural conditions (Figure 1D).

Figure 1. (A) Pieces of plant tissue (a leaf, for example) being fragmented to be placed on culture medium. (B) Clusters of shoots cultivated on a nutrient medium. (C) Flasks with proliferating shoots under controlled environment conditions. (D) Whole plant obtained through micropropagation. This plant is genetically identical to the plant from which it was originated.

Micropropagation offers advantages over traditional techniques for propagating plants, such as the mass production of uniform plants (clonal propagation), the possibility of producing plants without seeds, the elimination of viruses from plant tissues to obtain virus-free plants and the regeneration of genetically modified plants (Neumann; Kumar; Imani, 2009).

Production of transgenic plants

The advances in genetic engineering have made it possible to introduce foreign genes into cell and tissue culture systems. This has led to the development of genetically modified (GM) plants or transgenic crops with improved traits and characteristics.

All methods for genetically modifing a plant require a technique for introducing a foreign piece of DNA, conferring to a specific trait, into host plant tissue. Once inserted, the gene (termed the "transgene") should be incorporated into the host plant genome and stably inherited through future generations.

Although a number of different methods are available for introducing DNA into plant cells (Finer; Finer; Santarém, 1996; Santarém, 2000), particle bombardment and *Agrobacterium*-mediated transformation are the two methods that have proven to be the most efficient and are most commonly used for a large number of plant species (Finer et al., 2006). For particle bombardment, DNA is precipitated onto small (~1 μm) dense particles (either tungsten or gold) and accelerated towards the target plant tissue by a blast of helium. Once the particles enter the target cells, the DNA is released from the particles, becomes associated with the chromosomes and, if the proper conditions exist, the foreign DNA integrates into the chromosomes of the target cell (Finer; Finer; Santarém, 1996; Santarém; Finer, 1999; Trick et al., 1997). For *Agrobacterium*-mediated transformation, plant tissues are cultured with *Agrobacterium*, which is a bacterium that has the unique ability to introduce part of its DNA into plants. With the addition of signal compounds to the medium where *Agrobacterium* and the plant tissues are co-cultivated, and enhancing exposure of cells to the invading bacteria, the process of DNA transfer has become quite efficient for most plants (Santarém, 2000; Santarém et al., 1998; Trick et al., 1997).

Currently, genetic engineering in agriculture is focusing on genetic modifications to improve crop plants with respect to quality, nutritional value, and resistance to damage by pests and diseases (Table 1). Other aspects under investigation are how to increase plants' tolerance to extreme environmental conditions, to make plants better suited for their role in pollutant assimilation or to modify plants to produce materials which lead to the reduction of environmental pollution. In some cases, the intention is to reduce the amount of herbicide and pesticides, or other agricultural chemicals required to produce a given crop yield.

Table 1. Examples of plant products of biotechnology

Product	Trait
Alfalfa	Herbicide tolerance
Canola	Herbicide tolerance, modified fatty acid content, male sterility
Cotton	Herbicide tolerance, insect resistance
Flax, Linseed	Herbicide tolerance
Maize	Drought tolerance, herbicide tolerance, insect resistance, male sterility, fertility restored, altered amino acid content
Melon	Delayed ripening
Papaya	Virus resistance
Potato	Insect resistance, virus resistance, modified amylase
Rice	Herbicide tolerance, insect resistance
Soybean	Herbicide tolerance, insect resistance, modified fatty acid content
Squash	Virus resistance
Sugar Beet	Herbicide tolerance
Tomato	Delayed ripening, Insect resistance
Wheat	Herbicide tolerance

Source: ISAAA GM Approval Database. *http://www.isaaa.org/gmapprovaldatabase/default.asp*

Insect Resistance and Reduction of Pesticides

Plants have natural defense mechanisms that protect them from attack by insects, however the damage caused by pests is still sufficient to reduce the commercial potential of the crop. The traditional manner of protecting crops from pest damage is to spray the cultivated fields with insecticides, which is not only costly and laborious but also hazardous to farmers' health and to the biodiversity. In the effort to reduce the amount of chemical insecticides being used, plants are being genetically modified to have increased self-defense against pests. The most common example of genetically transformed plants for insect resistance is the introduction of gene coding for the *δ-endotoxin* of the bacterium, *Bacillus thuringiensis* (BT). The Bt protein has been used safely as an organic insect control agent for over 40 years. As a recent advance in biotechnology, Bt-transgenic plants (e.g., maize, cotton, and soybean) coding for insect resistance have allowed most farmers to reduce or abolish the use of insecticide when protecting plants from harmful pests, which can cause significant damage and yield loss in many areas (Karthikeyan et al., 2012). Bt- technology has been proven to be a chemical free and economically viable approach for insect pest control in plants (Sanahuja, 2011). Moreover, Bt maize also reduces toxin contamination that arises from fungal attack on the damaged grain.

China, India, Brazil, Argentine and South Africa are the five principal developing countries that grow biotech crops. In 2010, Brazil had 25.4 million hectares planted with biotech crops, while China had 3.5 million. However, in China, since the 1990's, the effort to improve biotechnology has had extraordinary rewards: 13 insect-resistant cotton varieties have been released (Xu; Bai, 2002). More than 1.1 million hectares of these insect-resistant cotton varieties were grown in 2001, significantly reducing pesticide use and the cost of cotton production (Huang et al., 2001). In 2012, 7.2 million small resource-poor farmers in China grew 4.0 million hectares of Bt cotton with an adoption rate of 80%, cultivating on average 0.5 hectare per farmer (James, 2012). Also in 2012, Brazil, for the fourth consecutive year, was the global leader in GM-growth, increasing its hectarage of biotech crops more than any other country – an impressive record increase of 6.3 million hectares, up 21% from 2011, reaching 36.6 million hectares (James, 2012).

Conventional agriculture has impacted significantly on the environment and biotechnology can be used to reduce the environmental footprint of agriculture. To date, progress has been made on reducing pesticides use; saving on fossil fuels; decreasing CO_2 emissions through no/less ploughing; and conserving soil and moisture by optimizing the practice of no-till farming through the application of herbicide tolerance. The accumulative reduction in pesticides for the 1996 to 2009 period was estimated at 393 million kilograms of active ingredient, which is equivalent to a 17.1% reduction in the associated environmental impact of pesticide use (James, 2010; Karthikeyan et al., 2012).

In conclusion, transgenic crops may have a central role in protecting the crop from its insect pests and have been used worldwide to control major pests. Hence, biotech crops are essential but will not, by themselves, solve all the problems in agriculture. Good farming practices such as rotation and resistance management should be practiced on biotech crops as they are for conventional crops.

References

AKTAR, W.; SENGUPTA, D; CHOWDHURY, A. Impact of pesticides use in agriculture: their benefits and hazards. Interdisc. *Toxicology*, 2, p.1-12, 2009.

FINER, J. J. et al. Monitoring gene expression in plant tissues. In: GUPTA, S. D.; IBARAK, Y. *Plant Tissue Culture Engineering*. The Netherlands: Springer, 2006.

FINER, J. J.; FINER, K. R.; SANTARÉM, E. R. Physical methods for plant cell transformation. In: MEYERS, R. A. (Org.). *Encyclopedia of Molecular Biology and Molecular Medicine*. The Netherlands: VCH Publishers, 1996. p.458-65.

GUPTA, S. D.; IBARAK, Y. *Plant Tissue Culture Engineering*. The Netherlands: Springer, 2006.

HUANG, J. et al. Plant biotechnology in China. *Science*, 295, p.674-7, 2001.

IGBEDIOH, S. O. Effects of agricultural pesticides on humans, animals and higher plants in developing countries. *Archives of Environmental Health*, 46, p.218, 1991.

JAMES, C. Global Status of Commercialized Biotech/GM Crops: 2010. *ISAAA Brief*, 42. ISAAA: Ithaca, NY, 2010.

JAMES, C. Global Status of Commercialized Biotech/GM Crops: 2010. *ISAAA Brief*, 44. ISAAA: Ithaca, NY, 2012.

KARTHIKEYAN, A. et al. Genetically Modified Crops: Insect Resistance. *Biotechnology*, 11, p.119-26, 2012.

NEUMANN, K.-H.; KUMAR, A.; IMANI, J. *Plant Cell and Tissue Culture*: a Tool in Biotechnology. Berlin: Principles and Practice, Springer-Verlag, 2009.

SANAHUJA, G. *Bacillus thuringiensis*: a century of research, development and commercial applications. *Plant Biotechnology Journal*, 9, p.283-300, 2011.

SANTARÉM, E. R. Métodos eficientes para a transformação de plantas. *Revista Ciência e Tecnologia*, 15, p.79-88, 2000.

SANTARÉM, E. R.; FINER, J. J. Transformation of soybean [*Glycine max* (L.) Merrill] using proliferative embryogenic tissue maintained on semi-solid medium. *In vitro Cellular and Developmental Biology Plant*, 35, p.451-5, 1999.

SANTARÉM, E. R. et al. Sonication-assisted *Agrobacterium* mediated transformation of soybean immature cotyledon: optimization of transient expression. *Plant Cell Reports*, 17, p.752-9, 1998.

TRICK, H. et al. Recent advances in soybean transformation. *Plant Tissue Culture and Biotechnology*, 3, p.9-26, 1997.

U. S. Geological Survey. 1999. *The quality of our nation's waters – nutrients and pesticides*. Circular 1225. Reston VA: USGS. Available at: *http://water.usgs.gov/pubs/circ/*
circ1225/.

XU, Z-H.; BAI, S-N. *Impact of biotechnology on agriculture in China*. Trends in Plant Science, 2002. DOI: 10.1016/S1360-1385(02)02305-1.

9

Environmental Preservation and Protection of the Built Environment in Megacities

Clovis Chiezzi Seriacopi Ferreira[1]

Introduction

Among the megacities that exist today (Castells, 2002), some have been built in less than one century. São Paulo may be one of the main examples of this. It went from being a small village with less than 100,000 habitants at the end of the 19th century to a city with more than 20 million at the end of the following century.

Although the science of planning was quite well-known at the time, the growth has been so fast that it seems the simple idea of planning has always come too late. Maybe, also because of the Portuguese culture, deeply rooted in real estate property, the common good never seems to be greatly considered. Planning properly, and mainly its consequent implementation, came only very later, only after the chaos was already built and consolidated.

Now we face a very uncomfortable dilemma. We have to preserve the environment, and, at the same time, we have to protect the built environment that we created within this megacity. To solve this

1 Universidade Paulista – UNIP, São Paulo, São Paulo, Brasil E-mail: *archiezzi@yahoo.com.br*.

dilemma we have to try and understand how we arrived at this point. Firstly, we have to look at the past and try to understand the evolution of the people that made this city and how this environment has been built, rather than cimply accusing people or proposing short- or long-term easy solutions.

Actually a little bit more or less than a century ago, São Paulo was only a very small village (Reis Filho apud Ferreira, 2006) with "a sleepy life in a cramped environment, with a few modest and poor houses... Until the mid-nineteenth century, São Paulo has not known the lordly residences that were so common and frequent in the Court at the time of the royal family..., and has not known magnificent mansions like those in Recôncavo Bahiano or even like those in Minas Gerais" (Lemos, 1978).

Figure 1. Downtown in 1904.

The coffee culture was mainly responsible for giving it a new way of life, transforming this small village in one of the most cosmopolitan megacities of the beginning of this third millennium (Lemos – Porto – Rolnik – Martins apud Ferreira, 2006). However, coffee alone cannot explain such development. The development of rail transport and the diversification of production and trade, import and export was also fundamental, which were introduced by immigrants, coming from Italy, Germany, Spain, Japan and several other countries.

But the education of children and young people was particularly essential, particularly that of the girls in schools of French or Italian origin, which greatly intensified after the arrival of those immigrants… (Chernow – Porto – Langenbuch – Rolnik – Lemos apud Ferreira, 2006). There were sons of farmers coming to live in São Paulo, from a new class of rich coffee producers, to study law, engineering or medicine, cohabiting with so many different cultures in such a small place.

These young men soon became interested in business, industry, banking, journalism and the arts. The arts, particularly theater and music, attracted farmers that, eased by railroad access, now can came more frequently to São Paulo. Theater and music gave rise to a new nightlife and, with so many different immigrant cultures, a new gastronomy. Before long the coffee producers found that they could keep their farm production even while living in São Paulo. New developments were built for these new classes, such as Higienópolis, Cerqueira Cesar with Avenida Paulista, and the firsts suburban areas, created by an English company, such Jardim América, Jardim Europa and City Lapa (Reis Filho, 1994). Because of these new habitants, a new gamma of services and business was created.

Manufacturing, banking, industry, trade, services, import, export, entertainment, culture. In few years, besides being an *earth of the drizzle*, São Paulo received another slogan: *the city that grows up most around the world.*

At the same time, probably for all these circumstances together, the Port of Santos soon became the busiest import and export route for trade in Brazil. Due to the development of the road transports, by the middle of the twentieth century, the main highways of the country begin to cross São Paulo.

And that is not the worst aspect: the main avenues and streets of the city become the natural connection among them! Suddenly, it seems that all the trucks, from and for all places in the country, must go through the City of São Paulo...

In this urban built environment, day and night, winter, spring, summer or fall, the day by day scenery became as shown at the next pictures.

Trucks carrying things that are not made in Sao Paulo and also that will not be used in São Paulo compose at least half of the traffic on the roads.

Figure 2. Main highways in Brazil cross the City of São Paulo.

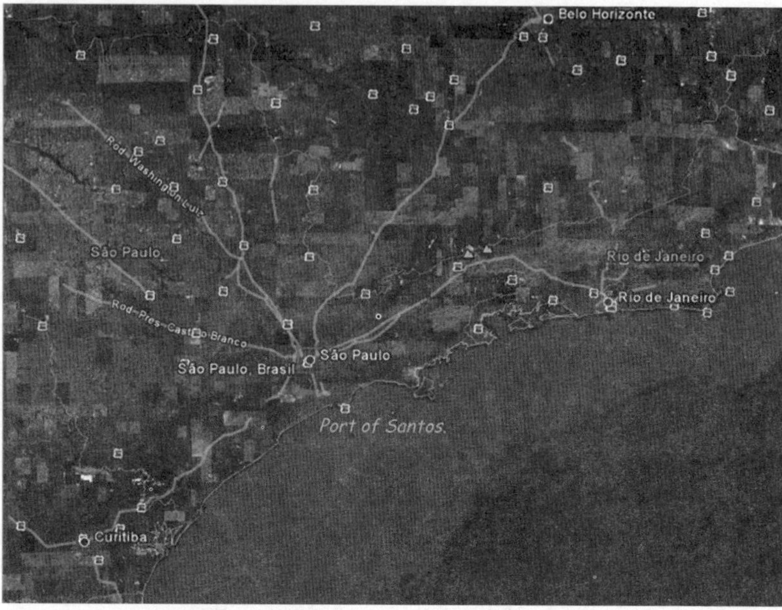

Figure 3. Day by day of avenues in São Paulo at different hours.

What can now be done to recover some quality of life for the people who live in such an urban environment? The simple and obvious solution is the same solution given for a lot of big cities around the developed world. A ring road around the city connecting the main highways that arrive in it.

However, like always seems to happen, São Paulo was late. Now there is a new kind of embarrassment. As shown in Figure 4, the regions surrounding the metropolitan area became Environmental Protection Areas.

Figure 4. Environmental Protection Area, in green, all around the Metropolitan Area, in red.

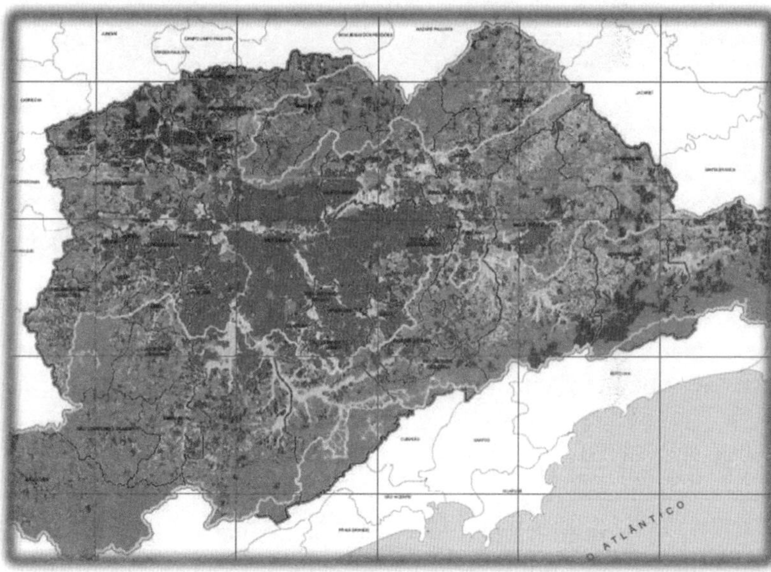

Although building a highway in an Environmental Protection Area is not as simple as it was thirty or forty years ago, as shown in Figure 5, the question for this case cannot be *what to do?* The real questions become: *How to do it?*

From an urbanism point of view, there is no other solution other than a ring road, at least not in the long-term. If Brazil had another transportation model... If they had planned the development better... If we change the matrix... If...

We can and have to change the matrix and models, but all alternatives would demand more than one generation to be implanted, even if it were possible, all the players involved would have to decide to work together and in the same direction.

Figure 5. Job site of the first lane of the highway *Imigrantes*, built across the *Mata Atlantica* in 1970s.

Now, what engineering can do is just do it causing as little damage as possible to the environment, even running the risk of being accused of being pragmatic.

Environmental Licensing

In Brazil, Environmental Licensing is intended to harmonize the protection of the environment with the need to carry out works or activities that, despite possibly having an impact on the environment, are necessary for social and economic development. According to Brazilian law, an Environmental License is an "administrative act by which the competent environmental agency establishes the conditions, restrictions, and environmental control measures that should be obeyed by the contractor, person or entity, to locate, install, enlarge and operate projects or activities using environmental resources considered effective or potentially polluting or those which, in any form, may cause environmental degradation" (Brasil, 2007).

The complete environmental licensing consists of a logical sequence of three authorizations that may be revoked at any time, whenever the entrepreneur does not meet the established conditions:

- Previous License (*Licença Prévia*)
- Installation License (*Licença de Instalação*)
- Operating License (*Licença de Operação*) Previous License is simply intended to verify whether it is possible to harmonize the work or activity with their environment. Once this enterprise is possible, the entrepreneur should prepare an Environmental Impact Study (EIA) and submit a Report of the Impact in the Environment (RIMA).

Installation License is a document authorizing the entrepreneur to begin construction. It means that all the details and specifications of the executive project meet the requirements established in order to minimize the Impact on the Environment. Operating License is the document authorizing the entrepreneur to start the activity or use the completed work. It certifies compliance with the requirements of the works and also establishes conditions for the exercise of such activities by a certain time. Before the end of this period, the entrepreneur must request a new license to operate, which may contain new requirements.

Under this context, there are two interesting recent cases in which technology had been largely applied to the benefit of the environment. The ring road around the metropolitan area of Sao Paulo and the new highway connecting this area to the Port of Santos. Both cases will be covered at the same time because, besides the cases having been undertaken almost at the same time, the solutions created in each one of them have followed basically the same principles.

Principles, Knowledge and Techniques Applied

First of all, among the conditions, restrictions, and environmental control measures established by the environmental agency to permit its construction, can be detached:

a) Both roads had to be Class 0 Highways, which means an express road, with no access or exit beyond the connection points. This condition was established to avoid unauthorized access to the Environmental Protection Area, mainly in order to prevent illegal real estate speculation in the forest.

b) The road would never divide the forest. So fauna can circulate freely at every single point, passing over the tunnels or under the viaducts, without ever needing to cross the road.

c) In case of accidents, any liquid that eventually leaks from vehicles is collected by special drainage system, in order that it does not reach the forest.

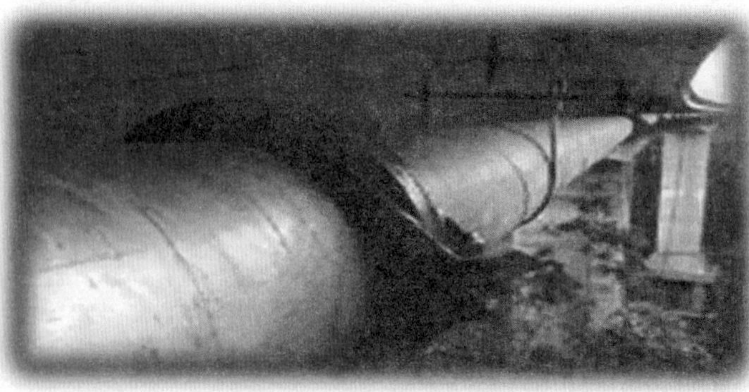

d) All the water used in the job site would be treated and reused. The eventual excess also had to be treated before being discharged.

A series of technical definitions used to minimize the damage to the forest are as follows.

a) The tracing of the new lane of the *Imigrantes* Highway was designed considering it being used as a service road to replace the old service road, built about 30 years previously, to support the building of the first lane. This way a new one and a large area was not needed so there was no need for further deforestation.

b) To avoid any contact of the water from the concrete with the groundwater, piles of foundation have to be made with the protection of pre-molded tubes applied before digging the soil.

c) The bases of the pillars were then made with minimal deforestation and protection so that there was no contact between the concrete water and the forest.

d) The pillars were molded with a formwork climbing system to ensure that no additional deforestation was required other than that done for the base.

e) The concrete of the roadway board was molded with a sliding formwork, in successive swings from each pillar, to avoid making contact with the forest.

f) Another construction system was adopted using an area not yet recovered, namely a remnant of the old construction site. Large slices of lane were built in one piece and then moved to their place, over previously made pillars, like a big train.

The most critical situation was a bridge over the *Billings Dam*, which provides clean water for almost half of the population of the metropolitan area.

In order to avoid absolutely any contact with its water, a factory was built to mold all the pieces of concrete.

The entire job site was isolated by a rock dam. Then all the pieces, except for the blocks and pillars, were brought to their places.

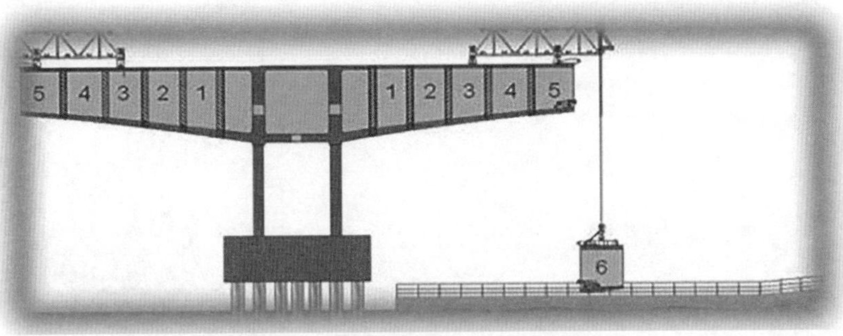
5
4
3
2
1
1
2
3
4
5
6

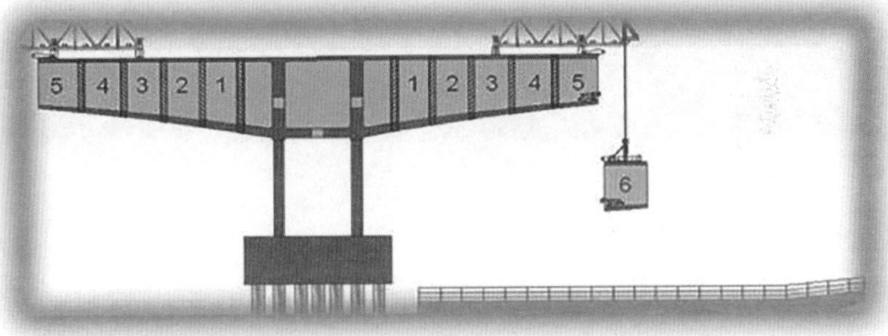
5
4
3
2
1
1
2
3
4
5
6

Conclusion

It is always possible to harmonize completely the protection of the environment with the need to carry out works or activities that, despite possibly having an impact on the environment, are necessary for social and economic development.

References

BRASIL. Tribunal de Contas da União. *Cartilha de Licenciamento Ambiental*. Brasília: TCU, 4ª Secretaria de Controle Externo, 2007.

CASTELLS, M. *Fim de milênio*. São Paulo: Paz e Terra, 2002.

FERREIRA, C. C. S. *A Casa dos Sonhos*: necessidades, aspirações, símbolos... São Paulo, 2006. 247f. Dissertação (Mestrado em Arquitetura e Urbanismo) – Universidade de São Paulo / Faculdade de Arquitetura e Urbanismo, São Paulo.

LEMOS, C. A. C. *Cozinha, etc.* Um estudo sobre as zonas de serviço da casa paulista. São Paulo: Perspectiva, 1978.

REIS FILHO, N. G. *Algumas experiências urbanísticas do início da República*: 1890-1920. São Paulo. Faculdade de Arquitetura e Urbanismo da Universidade de São Paulo, 1994.

10

How Information Technology Can Improve Quality of Life for Elderly People

Anita Maria da Rocha Fernandes[1]

Introduction

Two prominent trends are simultaneously occurring in our society: the aging of our population; and an increased reliance on computers (Zhou; Yokoi, 2010). The world´s population is growing older. The number of elderly people in the world will exceed the number of children under five (5) within ten (10) years, placing greater demands on a shrinking number of young caregivers and taxing social insurance programs, according to Waters (2009). According to the report "An Aging World: 2008," commissioned by the US National Institute on Aging (Kinsella; Wan, 2009), the number of people older than 5 will double to 14% from 7% of the world´s population in the next 30 years, rising to 14 billion by 2040 from about 506 million in the middle of last year. Beard et al. (2011) affirms that population aging is taking place in every country in the world. There are three factors underlying this trend: increased longevity (in most parts of the world, people are living significantly longer lives than in previous decades); declining fertility (as families

1 UNIVALI – University of Itajai Valley, Florianopolis, Santa Catarina, Brazil. E-mail: *anita.fernandes@univali.br*.

have fewer children, the proportion of the population that is elderly naturally increases); the aging of the "baby boomer" generations: the aging of large cohorts of children born after the 2nd World War in the USA – paralleled by similar booms elsewhere at various times – an area leading to high proportions of elderly people.

The most rapid rise in the elderly population is taking place in developing countries, where the increase in the number of people aged 65 and older is more than double the rate in developed nations. According to data from World Population Prospects: the 2015 Revision (United Nations, 2015), the number of older persons – those aged 60 years or over – has increased substantially in recent years in most countries and regions, and that growth is projected to accelerate in the coming decades. Between 2015 and 2030, the number of people in the world aged 60 years or over is projected to grow by 56 per cent, from 901 million to 1.4 billion, and by 2050, the global population of older persons is projected to more than double its size in 2015, reaching nearly 2.1 billion.

Besides the aging of our population, we have the advent of Information Technology. Nowadays, through advances in computerized technology and the internet, individuals are introduced to computers at a young age and become dependent on computerized resources for their daily lives. Most of this technology has come about fairly recently. An unfortunate consequence of this advancement is that the rapidity with which computers have taken over our lives has left one societal group lagging behind significantly in the area of computer literacy.

Currently, many elderly people are not able to use computers to their full potential and in many areas of many countries access to public computers is very limited. Despite this lack of ability and access, considerable information about programs and services available to assist the elderly are now provided online. The internet is becoming an increasingly significant means by which friends and family members can keep in contact with each other. Keeping in regular touch with friends and family, shopping from home, accessing high quality entertainment, and having home care, are goals which elderly people are likely to want to achieve. According to Alm, Gregor and Newell (2002), some examples of special ways in which IT could be developed, which would be of great benefit to elderly people, are:

- Communication and social connectivity: with the breaking up of extended family networks which exist physically close to each other, loneliness and social isolation are increasing problems faced by many elderly people and their families. Current technology such as email and video-telephony can help to bridge this communication gap. The development of more ambitious forms of communication and contact at a distance will further help to keep elderly people in touch with their families and friends;
- Access to information and services: being able to shop and access services from home is clearly an advantage for people who have difficulty getting out. If such services are too difficult to be used by elderly people, they will have to be made easy for use by a generation who have not been brought up with computers. With such a widening of their functionality, a large new market could open up for providers of goods and services;
- Promoting lifelong services: the development of computer-delivered education and training could benefit elderly people who want to keep their minds active. In addition to the direct educational benefit and enjoyment, taking on new learning challenges is believed to be one way of keeping mental faculties preserved;
- Telecare and Telemedicine: the increasing cost of medical care for an aging population is an important concern. It is possible that providing remote access to a range of services could help alleviate this problem, and could play a part in encouraging a self-help approach to staying healthy;
- Remaining economically active and productive: an economic concern about the aging population is that a large number of retired people must be supported by shrinking number of working taxpayers. One solution for this is to allow and encourage people to remain economically active beyond the usual retirement age. IT offers possibilities for elderly people to accomplish this more easily, through opening up new information handling job opportunities and allowing people to work from home.

As our society has advanced to one in which technology has been mainstreamed and the use of computers has begun to address the psychosocial value of computer use, especially as it relates to enhancing the quality of life for elderly people.

According to Hutchison (2010) and Keegan (2008), access to information and computer technology can help negate aspects of isolation by inviting new opportunities for learning, communication, and socialization via the internet and the web. Numerous studies have evaluated the effects of activity, learning, personal goals, and continuing self development (all possible outcomes of computer use), and have found that levels of life satisfaction and meaning are enhanced and positively correlated with increasing feelings of control, self-esteem and self-efficacy. Ruff and Keyes (1995) identified several criteria of psychological well-being: autonomy; feelings of life; positive relations with others; environment; purpose of life; self acceptance; and personal growth.

Lapierre, Bouffaid and Bastin (1997) reported that many elderly individuals believe that continued self-development is as important as leisure activity and that by pursuing individual aspirations, needs for well-being, meaning in life, and continued contribution to society are met. McConatha et al. (2004) explain that using the internet with others may help reduce loneliness and the rate of suicide among seniors. Even in terms of treating ailments such as arthritis, spending time online and creating an active mind may reduce the amount of pain experienced and the amount of medication needed. White et al. (2002) found that older adults use computers to communicate, explore interests and hobbies, or obtain consumer information and access community resources, thus increasing their autonomy. All these studies revealed many important things: elderly people believe they will benefit more from personal instruction in using computers, rather than from manuals. Tutorials can be useful if they are "step by step"; elderly people identified young persons in their communities (their grandchildren, school children taking computer classes, scouting youth, etc.) who might prove to be valuable allies in helping them use computers. Finding other volunteer tutors might be possible through media sources in communities; while some elderly people already appreciate the value of computers, others are skeptical that they would benefit from using one; they continue to prefer newspapers or other sources for information; interest in e-mail voiced by many of the elderly people as a vehicle for enhancing their contact with their children and grandchildren (almost all of whom own a computer). In fact, most of the elderly people who owned computers had been given them by family members; and physical limitations, notably eyesight problems and arthritis in the hands, are barriers for a sizable portion of elderly people.

Considering these aspects, here we present some examples about how IT can help elderly people and improve their quality of life. In the next sections we will present aspects related to social inclusion, tools to help elderly people use computers, and home care.

Elderly People Social Inclusion By It

According to Morrison and Barnett (2008), social isolation and feelings of loneliness are common problems for elderly people. Recent research has demonstrated a worrying trend that many who work with elderly people have observed themselves: fear of being alone is a mayor source of anxiety as people grow older; some elderly people go for days without seeing another person; some die alone; 16% of elderly people in deprived inner city areas suffer severe loneliness; certain ethnic groups are particularly vulnerable; isolation and loneliness can damage both mental and physical health in elderly people. In this context, how information technology can help prevent and alleviate isolation and loneliness amongst elderly people is an important theme to study.

Over the last decade, an array of technologies have changed the way many of us interact. Internet communication systems such as e-mail and social networking sites like Facebook and Twitter have revolutionized personal communication. The use of technology by elderly people brings potential social benefits such as reduced isolation and increased social participation. IT is not merely an end in itself, but it can be a means to enable elderly people to renew and develop social contacts and engage actively in their communities. According to Morrison and Barnett (2008), it can provide opportunities to: participate in meaningful work and other activities (whether paid or on a volunteer basis); interact in new ways with family and friends; learn, develop skills and gather experience; share learning, skills and experience with others. According to Allen (2008), some of the important barriers to older people adopting digital technologies include: lack of access to the internet at home: only half of people aged 60-79 have access to the internet at home, but this falls to 17%

among the over 70s. Adults over the age of 60 are also less likely than younger adults to have internet access. The dominant reason for not having or seeking access is that elderly people do not feel that they need it; low awareness of what technology can offer: 10% of people aged 60-69 have access to the internet but do not use it. They feel that digital technology has no relevance for them and that they would gain nothing by using it; inadequate marketing: technology marketing is generally aimed at the young, promoting gimmick aspects of products that do not interest elderly people. Or, marketing is aimed at the frail elderly, a group with which most elderly people do not identify; inappropriate design: digital equipment is designed to attract young buyers who have grown up using technology. Small buttons, fiddly controls and unnecessarily complicated interfaces can all be barriers to the elderly, or less adept, users. The appearance of "special" equipment is also a deterrent for some elderly people who do not want ugly objects cluttering up their homes; anxieties: elderly people tend to have certain fears regarding technology. One of them is cost: they assume, for example, that computers cost more than they actually do. Another is breaking equipment or doing something wrong. A third is security: although most elderly people do not know enough about technology to be familiar with common security problems, many know enough to be concerned.

Tools to Help Elderly People to Use Computers

This topic contains some examples about tools that can help elderly people use computers such as software that simplifies the interface, keyboards with extra-large keys, voice recognition device, and screen resolution setup.

According to Cintrón (2011), one reason why some elderly people find computers difficult to use is the sheer complexity of the Windows interface. For this reason, many third-party application programs are developed and designed to simplify how items and functions appear on the computer screen. For example, the home screen with Pointer-Ware, software developed specifically for the elderly, only appears as a simplified white background with several enlarged icon buttons that directly connect to the most commonly used features. You can also set which items to include on the home screen such as mail, photos, internet, games, etc. Depending on the developer, software for the elderly can feature many other functions designed to simplify computer use.

With old age, many people start losing control over their motor skills, and simple activities such as typing can become extremely challenging. Some also develop severe finger arthritis. With these elderly people in mind, many computer hardware companies started developing special keyboards with extra-large keys (Touchbrite, 2013). Some even feature 1-inch-sized keys, such as Vision 2 Large Keyboard. And for those with impaired eyesight, many of these keyboards are also designed to make each character easily identifiable. This is achieved by employing highly contrasting colors. For example, some keyboards for the elderly developed by Keys-U-See are designed with bright yellow keys which contrast with the characters in dark print.

Another solution that can help elderly people with their computers is the use of voice recognition programs. These applications are designed to relay commands to the computer and make it perform tasks without the use of the keyboard, mouse or other input means. Simply by saying the keywords, your computer can open applications and documents, switch between programs and also close or stop applications. Some are also designed to transcribe voice into text, which can make writing e-mails a lot easier (Hope, 2013). The Windows Vista Operating System, for example, induces a voice recognition application. To make these programs work, a microphone is needed.

Sometimes, however, tools that are readily available in Windows and Macintosh are enough to make computer use easier for the elderly, specifically if the problem only has to do with poor eyesight (Cintrón, 2011). A simple tweak of the screen resolution to enlarge the icons and can make the color scheme less complicated will do the trick. The settings for this can be found in the properties section. As for elderly people with weakened motor skills, a mouse with a highly sensitive response setting can prove difficult to control. This setup can be changed in devices and printers section by right clicking the mouse icon and going to the mouse settings option.

Home Care For Elderly People

Here we show some examples about automated intelligent houses, ambient emergency detection, intelligent bathrooms, and robots to assist elderly people.

Automated "Intelligent" houses help elderly people stay in homes longer

In the future, automated homes, intelligent environments, modern sensor systems, and information technology can help elderly live autonomously in their own homes.

These systems remind residents to regularly take their medicine, for example, or send out an alert if someone falls. The technology remains discreetly in the background – until it is needed (Rauber, 2003). Researchers at the Fraunhofer Institute for Experimental Software Engineering IESE in Kaiserslautern, Germany, are creating these intelligent environments. These scientists are developing information technology concepts for a system that collects detailed environment information via a network of numerous, unobtrusively installed sensors that analyze and respond to specific situations. Using many hidden sensors, the system monitors the daily routine of the occupants (Fraunhofer-Gesellschaft, 2009).

Ambient emergency detection – safety for autonomous living

One particular challenge is to develop a system that not only functions safely and reliably, but also continuously adapts to changing requirements. New assistance services and devices can easily be integrated into the system, such as motion detectors or pressure sensors in the mattress (Fraunhofer-Gesellschaft, 2009).

These sensors automatically report their data via wireless communication to a control center concealed in a cabinet. If so desired, the sensor's functions can be manually controlled by a computer screen on the nightstand.

Should users prefer not to concern themselves with the technology, they can leave everything to the system, which operates autonomously. The system can also detect whether an occupant has fallen, communicate this information via telephone or internet to a pre-selected contact person. This many be a relative, a neighbor, a care facility, or an emergency medical center (Rabe, 2009).

The Bathroom, your friend and helper

The "Assistive Bathroom Environment" from the Fraunhofer Institute for Microelectronic Circuits and Systems (IMS) in Duisburg demonstrates how these support services function. The obstacle-free electronically equipped bathroom has a toilet that recognizes the user and automatically adjusts itself to a suitable height. A light automatically switches itself on and off, and the tap shuts itself off to save water (Bock; Georgoulas; Linner, 2012). This bathroom even has the capability to monitor the amount of toothpaste being used. The goal is to monitor and support the occupant without being intrusive. The new cutting-edge bathroom is particularly helpful for elderly and infirm people who are sometimes a little disorientated and may confuse important everyday tasks. Illuminated pictograms in the mirror indicate what there is for the user to do next: for instance, wash, brush his or her teeth and shave – even the days for showering are permanently saved. A recorded female voice gives the bathroom's occupant a friendly reminder of when it's time to take his or her medication (Rabe, 2009). The individually tailor-made support assistance is combined with the documentation of the procedures that occur in the bathroom. Sensors in the door, toilet, tap, light switch, and carpet detect every activity and record them electronically. Above all, this is important if the user eventually needs professional care. Doctors or care personnels can see from the computer records what personal hygiene tasks the person under care has completed, how often he or she has visited the bathroom, used the toilet, or whether he or she has fallen down. In case of an emergency, the computer automatically alerts the chosen contact person or calls the care center (Rabe, 2009).

Domestic robot to help sick elderly live independently longer

To enable elderly people to live at home as long as possible, a group of European researchers, coordinated from the Eindhoven University of Technology, will link robots and "smart homes" (Eindhoven University of Technology, 2010). The robot, a "sensible family friend," will ensure that the home is a nice place to stay and that patients do the right things.

Researchers at the University of the West of England, Bristol (UWE) are taking part in a European Project aimed at creating an intelligent system comprising a robot and smart sensors that can support independent living for elderly people (University of the West of England, 2010). This project will bring together a multidisciplinary team at the UWE with expertise in care for elderly people: robotics, human computer interaction, a team to look at the needs of elderly people and the potential of technology to meet those needs with companies such as Robosoft (France), Smartex (Italy), CSEM (Switzerland) and Smart Homes (Netherlands). The team will also include health care providers from St. Anna hospital in the Netherlands and the universities of Thessaloniki and Lappeenranta. The project aims to produce three key care systems for elderly people. A wearable health status monitor with smart sensors woven into undergarments (Smartex and CSEM), a secure tele-alarm and health reporting system, and a nutrition support system (Thessaloniki), which might, for example, consist of reminders for when meals and drinks should be taken.

Conclusions

This paper wanted to show how IT can help elderly people have a better quality of life. Many solutions are available, but our society does not know that it is possible to use these technologies. Another problem is that the society is not prepare to use them, because in many countries, elderly people do not have access to computers. Thus, it is necessary that governments create policies to make IT available to elderly people.

It is necessary to stimulate older people to use computers and the internet, it is also necessary that cities make the infra-structure to use IT available to these individuals.

References

ALLEN, J. *Older People and Well Being*. Institute for Public Policy Research, 2008.

ALM, N.; GREGOR, P.; NEWELL, A. F. Older people and information technology are ideal partners. *International Conference for Universal Design*, Yokoama, Japan, Nov./Dec., 2002.

BEARD, J. R. et al. *Global Population Ageing*: Peril or Promise. Geneva: World Economic Forum, 2011.

BOCK, T.; GEORGOULAS, C.; LINNER, T. Towards robotic assisted hygienic services: concept for assisting and automating daily activities in the bathroom. *Gerontechonoly*, v.11, n.2, 2012.

CINTRÓN, F. *Seniorama Review*: best guide to help the elderly use computers. Self Growth.com, Dec., 2011. Retrieved January 2, 2013. Available at: http://www.selfgrowth.com/articles/seniorama-review-best-guide-to-help-the-elderly-use-computers.

EINDHOVEN University of Technology. 2010. Domestic robot to help sick elderly live independently longer. *Science Daily*. Retrieved January 5, 2013. Available at: http://www.sciencedaily.com/releases/2010/04/100422085218.html.

FRAUNHOFER-Gesellschaft. 2009. Automated Intelligent Houses Help Elderly Stay in Homes Longer. *Science Daily*. Retrieved January 5, 2013 Available at: http://www.sciencedaily.com/releases/2009/02/090213115332.htm.

HOPE. *Assistive technology for elderly people*. Retrieved January 9, 2013. Available at: www.utdallas.edu/~chung/Hope/index.html.

HUTCHISON, P. Elderly living alone use internet to keep in touch with family. *The Telegraf*. September 22, 2010. Retrieved January 5, 2013. Available at: http://www.telegraph.co.uk/technology/internet/8018203/Elderly-living-alone-use-internet-to-keep-in-touch-with-family.html.

KEEGAN, V. Connecting older people to the world. *The Guardian*. October 9, 2008. Retrieved January 2, 2013. Available at: http://www.guardian.co.uk/technology/2008/oct/09/mobilephones.computing.

KINSELLA, K.; WAN, H. An aging world: 2008. US Censo Bureau. *International Population Reports*, P95/090-1. Washington, DC.: Government Printing Office, 2009.

LAPIERRE, S.; BOUFFAID, L.; BASTIN, E. Personal goals and subjective well being in later life. *International Journal of Aging and Human Development*, 45, 4, p.287-303, 1997.

MCCONATHA, J. M. et al. Turkish and US attitudes to toward aging. *Educational Gerontology*, 30, p.169-183, 2004.

MORRISON, J.; BARNETT, A. *Older people, technology and community*: the potential of technology to help older people renew or develop social contacts and to actively engage in their communities. The Royal United Kingdom Beneficent Association, 2008.

RABE, A. 2009. *Intelligent Home Enviroment*. Retrieved January 5, 2013. Available at: http://www.fraunhofer.de/en/press/research-news/2009/02/Pressinformation5Februar2009.html.

RAUBER, M. 2003. Smart Houses may help seniors stay at home. *Media General News Service*. Retrieved January 5, 2013. Available at: www.marc.med.virginia.edu/news-smarthouse.html.

RUFF, C. D.; KEYES, C. L. M. The structure of psychological well being revisited. *Journal of Personality and social psychology*, v.69(4), Oct., p.719-27, 1995.

TOUCHBRITE. *Keyboards with large buttons*. Retrieved January 10, 2013. Available at: www.touchbrite.com/easy-use/keyboards-large-buttons.

UNITED NATIONS. 2015. Population 2030: Demographic challenges and opportunities for sustainable development planning. Available from http://www.un.org/en/development/desa/population/publications/pdf/trends/Population2030.pdf.

UNIVERSITY of the West of England. 2010. Could robots and smart devices help older people look after themselves? *Science Daily*. Retrieved 5, January 5, 2013. Available at: http://www.sciencedaily.com/releases/2010/03/100324184588.htm.

WHITE, H. et al. A randomized controlled trial of the psychosocial impact of providing internet training and access to older adults. *Aging & Mental Health*, 6 (3), p.213-21, 2002.

ZHOU, W.; YOKOI, S. ECSCHOOL Model: a case study of building learning support community in ICT promotion project in Japan. *International Journal of Cyber Society and Education*, n.53-68, v.3, n.1, June, 2010.

11

Poverty in metropolitan areas and conditional cash transfer programs in Brazil: the Campinas (SP) experience

André Pires[1]

Brazil and China are more similar to each other than one might imagine. Despite the differences in their individual history, cultural characteristics, political and social institutions – as well as the physical distance that separates them – there is a group of socioeconomic affinities and common challenges that are worth highlighting.

Chart 1. Selected socio economic indicators – Brazil and China (2010)

	China	Brazil
Population	1,330,045,000	191,909,000
Territory (km²)	9,596,960	8,547,403
GDP (USD million)	5,878,000	2,090,000
GDP per capita	4,419.40	10,890.58
HDI	0,663	0,699
Life expectancy at birth (years)	73.5	72.9
Expected Years of children schooling (years)	11.4	13.8

Source: Human Development Report 2010. New York, United Nations Development Program, 2010 (http://hdr.undp.org/en/reports/global/hdr2010).

---

1 PhD. Professor of the Graduate Program in Education, Campinas Catholic University (PUC-Campinas).

Firstly, both countries are populous and cover wide territories. China, in addition to having the largest population in the world, it also possesses the third largest territory. Brazil, on the other hand, has a population that is around seven times smaller than China and is the fifth most populous country in the world, with the fifth largest territory. These common characteristics allow both countries to build their history upon their natural resources, and especially upon their great cultural diversity – to the point of being respectively the seventh and the second largest economies in the world. Anthropologist Claude Lévi-Strauss (Lévi-Strauss, 1987) has already taught us that the success of cultures that have achieved more cumulative forms of history is not a result of some supposed "isolated superior culture," but of the exchange and combination of voluntary and involuntary processes in which cultural diversity plays a fundamental role. It is important to mention that these contributions have not always been recognized by both countries, and the current ethnical conflicts that exist both in Brazil and China should remind us of this history.

Brazil and China are also facing common challenges. In fact, the economic growth experienced by these countries has not yet been translated into a more equanimous distribution of resources and guarantee of access to health and basic education public services for the population. Table 1 suggests that despite the economic strength of both countries, Brazil and China were in seventy-third and eighty-ninth position in the Human Development Index (HDI) in 2010, which takes not only the country's Gross Domestic Product but also health and education indicators into consideration. Life expectancy at birth in these two countries is around 73 years, which places them far from the indicators of countries with a high human development. Similarly, the expectancy for time spent in education by Brazilian and Chinese children is 13.8 and 11.4 years, respectively, which is also less than the countries classified by the UN as having a high human development.

One could further mention many other common challenges to be faced, but this article will focus on the matter of combating poverty. To do so, I will consider the Brazilian experience of cash transfer policies, specifically the "*Bolsa Família*" (Family Grant) Program, created by the Federal Government in 2003, which today serves more than 45 million people.

This chapter will be divided in three sections, not including the introduction. In the first section, I will introduce some information about how poverty in Southeast of Brazil is concentrated in metropolitan areas. Subsequently, the "*Bolsa Família*" Program (BFP) will be detailed and the third part will include information of the results from a research project conducted with the PBF beneficiaries that live in the Municipality of Campinas (SP). This article aims to show that the Cash Transfer Program is extremely important, albeit insufficient, for creating conditions so people can lift themselves out of poverty. As these pages will demonstrate, the fight against poverty should combine income transfer policies and other social policies in the areas of health and education.

Extreme poverty in Brazil, according to Census 2010 data

In 2011, the Brazilian Ministry of Social Development and Fight against Hunger (MSD), upon publicizing the "Brazil with no Misery" Plan, which articulates a group of federal government actions, including the BFP, aimed to fight the problem of extreme poverty and introduced a characterization of the Brazilian population that live in these conditions based on information obtained during the 2010 Demographic Census. The MDS defined a person as being in a situation of extreme poverty as one whose monthly household income per capita is equal to or less than to seventy reais (R$ 70.00).[2] The poverty line defines who is eligible to participate in the PBF, as we will see below, this is a monthly household income per capita equal to or less than one hundred and forty reais (R$ 140.00).[3] Some of the information gleaned from the 2010 Demographic Census is presented below.

Chart 2. Brazilian population per region (2010)

	Total Population	% Total	Extreme Poverty	% Total
North	15,864,454	8.3	2,658.,452	16.3

2 Approximately US$ 40 in 2011.

3 Approximately US$ 90 in 2011.

	Total Population	% Total	Extreme Poverty	% Total
Northeast	53,081,950	27.8	9,609,803	59.1
Southeast	80,364,410	42.1	2,725,532	16.8
South	27,386,891	14.4	715,961	4.4
Central-West	14,058,094	7.4	557,449	3.4
Brazil	190,755,799	100.0	16,267,197	100.0

Source: IBGE demographiccensus 2010

Chart 2 suggests that the Southeast is the most populated region in Brazil (42.1% total) followed by the Northeast (27.8%). Both regions together account for 70.0% of the Brazilian population. However, when taking the extremely poor into account, the relative weighting of the Southeast (16.8%) decreases, significantly increasing the participation of the Northeast (59.1%). Therefore, the distribution of extreme poverty per region in Brazil is uneven, with it being mostly concentrated in the northeast and southeast regions, which represent around 76.0% of all the extremely poor people in the country.

When comparing the population in the northeast and southeast regarding where they live, we will also see different configurations of extreme poverty, as is depicted in the chart below.

Chart 3. Distribution of the population in the Southeast and Northeast, according to location of residence (2010 in %)

	Urban	Rural	Urban (e. poverty)	Rural (e. poverty)
Northeast	73.1	26.9	47.5	52.5
Southeast	92.9	7.1	78.7	21.3

Source: IBGE demographic census 2010.

In terms of place of residence, it is noticeable that the majority of people in both regions live in areas classified as urban, with a higher concentration in the Southeast. In the Northeast, the extreme poor mostly live in rural areas (52.5%), whereas in the Southeast the majority of extreme poor live in urban areas (78.7%). This is an important piece of information since Campinas (SP), from where this article's information was collected, is located in the southeast of Brazil and is central to a Metropolitan Region that groups nineteen municipalities with total population of 2.6 million people.[4]

"Bolsa Família" Program

The "Bolsa Família" Program (PBF), a federal government conditional cash transfer program, was implemented in 2003 with the goal of fighting poverty by articulating actions over two time periods.[5] In the short term, the money transfer to poor families had the goal of alleviating the immediate effects of poverty, providing conditions to acquire basic goods and services for their living. In the long term, there are requirements that families must fulfill in order to receive the benefit, noticeably those associated to education and health, the purpose of which is to fight the mechanisms that reproduce poverty. In education, "*all children and adolescents between 6 and 15 years old must be duly enrolled in a school and have a minimum attendance of 85%. Students between 16 and 17 years of age must have minimum attendance of 75%*" (Ministry of Social Development and Fight against Hunger, 2011). In regards to health, in order for families to be eligible for the benefit, they must commit to following the vaccination card system and monitoring the development of children younger than 7 years of age. Women between 14 and 44 years of age must also be monitored, and pregnant or breastfeeding women must undergo prenatal testing and monitor theirs the baby's health. In the long term, the BFP should also be articulated with other public policies, namely health, education, sanitation, housing, among others.

The BFP is targeted towards poor families which are defined by the government as families with a monthly income of less than US$ 90 per capita. It transfers, per family, benefits from US$ 40 up to US$ 160, according to family composition and poverty level. The basic

4 According to the Census 2010, the population in Campinas is 1,080,113 inhabitants, which represents approximately 40% of the total population of the 19 municipalities that compose the Campinas Metropolitan Region (RMC).

5 Analyzing the historical conditions that made the creation of PBF possible through Law 10936/2004 as well as its intrinsic characteristic of fighting poverty, is not our objective. For more information on this matter, see Cohn, 2010 and Silva; Yazbek; Di Giovanni, 2007.

benefit (US$ 40) is paid to families with a monthly per capita income of less than US$ 40), regardless of the number of children or adolescents. Variable benefit is paid as US$ 24 per children or adolescents until they are 17 years of age (maximum of 5 benefits per family), The benefits are paid by the government directly to the families, who can withdraw them on a monthly basis from the banking system or partner establishments by means of a cash card.

In the article "Profile of Beneficiary Families of the Bolsa Família Program," published in March of 2007 by the Ministry of Social Development and Fight against Hunger, it is notable that the direct income transfer to the poorest has had positive effects. As several studies demonstrate, (Soares et al., 2007; Soares et al., 2006) the pecuniary amount paid by the Program has contributed to reducing poverty in Brazil over the past few years. This has provided millions with a minimum income to ensure their survival.

If this short-term dimension is being served by the program, then the long-term actions deserve a more in-depth study. The program strengthens the right of poor families to access quality education and health services and makes integration between the "Bolsa Familia" Program and other public policies possible, whether they be at federal, state or municipal levels. Therefore, the next part of this chapter will reflect on the importance of combining the Cash Transfer Program with other public policies to help the poorest families leave their current vulnerable situation. In order to do so, socioeconomic information regarding the families served by the "Bolsa Família" Program in Campinas (SP) – present in the Unified Record of Federal Program Beneficiaries – provided by the City Hall through the Secretariat of Citizenship, Work, Social Assistance and Social Inclusion (SMCTAIS), will be used. These records contain information about approximately 16,500 households which are home to around 72,500 people.[6]

---

6 There are three cash transfer programs in Campinas: the City Hall's Minimum Family Grant Program (PGRFM), the State program Citizen Income Program and the Federal Government's "Bolsa Família" Program. Of these, the "Bolsa Família" is the one that reaches a greater number of families. In 2005, ¾ of all families that were part of the cash transfer programs in Campinas were part of BFP.

Income and Vulnerability

Below shows information on the composition of families in a situation of vulnerability regarding their income. Chart 4 shows that schemes for nuclear families represent most of the beneficiaries (55.6%). This group, comprising families made up by spouses and children (independent of number) represent almost half the total number of families analyzed (48.0%). It is important to point out that monoparental families – albeit less numerous – aggregate the expressive contingent of 36.6% of the total number of families.

Chart 4. Distribution of families according to the type of family arrangement (in %)

	%
Nuclear – Father, mother and children	48.7
Enhanced nuclear – Father, mother, children and relatives and/or others	6.8
Mono parental – Father or mother and children	27.2
Enhanced mono parental – Father or mother, children and relatives and/or others	9.4
Unclassified	7.8
Total	100.00

Source: Federal Government Unified Record.

We will now look at the pattern of income of families enrolled in the "Bolsa Família" Program.[7] Monoparental families, be they simple or enhanced, have a much lower have average income compared with nuclear families. The average income for monoparental families is R$ 287.90 (according to Chart 10), which is 27% less than the average income of nuclear families (R$ 393.10). When we take into consideration that the presence of children is constant in the studied schemes, we will see that monoparental families are more vulnerable in terms of income when compared to bi-parental families. This is because

7 Family income amounts were deflated by the INPC (National Consumer Price Index) calculated by IBGE (Brazilian Institute of Geography and Statistics), the most used index for this sort of procedure (Corseuil; Foguel, 2002). The deflation base is the year 2006.

the burden of raising children becomes a full responsibility for one of the parents, which often makes earning resources more difficult. It is also important to stress that families with parents and/or others (designated as enhanced) have a higher average income than those that only count the parents and children or one of the spouses and children.

Chart 5. Distribution of monthly family income by type of family according to the sex of the person responsible for the benefit (in R$)

	Male Family Income	Female Family Income
Nuclear	391.90	364.70
Enhanced nuclear	442.20	418.20
Mono parental	220.10	230.10
Enhanced monoparental	418.50	342.80

Source: Federal Government Unified Record.

Families in which the women is head tend to have a lower income than those headed by men in almost all the family arrangements studied, the exception being the monoparental families. In fact, the explanation for this discrepancy becomes intelligible when cultural factors, which establish different roles between gender, are taken into account. These assume that women must take care of both the children and the house. Actually, it should be stressed that the overwhelming majority of the analyzed households are headed by women (94.1%), which makes this fact even more problematic.

We found an apparent exception to the rule: monoparental families. In these cases, families headed by women have a higher income (approximately an extra R$ 10) compared to those headed by men. This is an interesting situation, as where men head households with children by themselves, they tend to create a less favorable combination between caring for their children and earning income. For women, partially because of the effects of many years of socialization between the gender roles, this task may be better executed.

Below there is data presented in the chart to show the impact of raising small children along with the family income composition.

Chart 6. Distribution of monthly family income by type of family with children of up to 6 years age (in R$)

	Children up to 6 years of age Family Income	Children up to 6 years of age in day care Family Income
Nuclear	346.90	424.50
Enhanced nuclear	403.00	536.40
Mono parental	222.20	167.00
Enhanced monoparental	304.30	335.00

Source: Federal Government Unified Record.

The presence of small children (up to 6 years of age) is certainly a contributing factor towards a decrease in family income. We know that children of this age require a supplementary care effort from the parents, which frequently makes it impossible for these people to hold remunerated jobs. The number analysis shows that families with children of up to 6 years of age have, in all of family arrangements under study, a lower average income compared with the total of the families.

It becomes important to show that public policy intervention can be an important factor to reverse this economic vulnerability situation. When we compare the average income of families with children of up to 6 years of age with the average income of families with children of up to 6 years of age in daycare, we can see that the former generally have a higher income. For nuclear families, the advantage for those who can put their children in daycare is about 22%. Among the enrolled enhanced nuclear families, it is 33%, and enhanced monoparental families, 10%. There is an exception for simple monoparental families, which present a decrease of 25% in their income. One hypothesis to understand this may be related to the daycare working hours in Campinas. Daycare working hours are normally between 7 a.m. and 5 p.m.; however, for various reasons, many facilities only open at 8 a.m. and close at 4:45 p.m. The need to leave the children in daycare later and pick them up earlier sometimes makes it impossible for these mothers to engage in paid employment. In addition, many mothers have more than one child, who sometimes are even too young to attend daycare. Despite the discrepancy with monoparental families, we can conclude that putting small children in daycare, thus avoiding private and individual solutions, can create fundamental economic security for women, and is, in many cases, the only alternative of care for their children.

Based on the above, we will now study the effect of retirement and pensions on the average income of these families. When analyzing the data from the chart below, we will see that the presence of people entitled to retirement and/or pension radically reduces the probability of the family being in a condition of vulnerability. The numbers are expressive in this sense. A nuclear family with a retired member or pensioner has an average monthly income of R$ 454.90, which is 24.0% higher than the average income of nuclear families (according to chart 7). For enhanced nuclear families, the difference in favor of those with retired people/pensioners increases to 43.0%. For monoparental families, when there is a retired person/pensioner, the average income is approximately R$ 295.80, which is 29.0% more than their average income. Finally, the difference in favor of enhanced monoparental families with the presence of retired/pensioners is 30%.

Chart 7. Distribution of monthly family income per capita, by type of family where there is a retired person/pensioner (in R$)

	Family Income
Nuclear	454.90
Enhanced nuclear	601.80
Nuclear (average)	528.40
Monoparental	295.80
Enhanced monoparental	451.60
Monoparental (average)	373.70

Source: Federal Government Unified Record.

It is important to point out that, as seen with daycare, the importance of the Government role, through their public policies, of reversing the situation of vulnerability for these families. As the Campinas example suggests, social care policies, such as the universalization of the right to retirement, are fundamental for the social reproduction of a number of families as these families can earn the minimum wage with it, which, despite not being high, is incredibly significant for their situation, and is also a lifelong benefit.

Lastly, it is important to point out an additional factor that helps us understand the socioeconomic situation of families enrolled in the system – namely the uneven distribution of public goods and services within a certain territory. It is a true challenge for social policies to break what is known as "*the place effect*," to coin Pierre Bourdieu's expression (Bourdieu, 2003), which exists in areas with a great concentration of poor people and makes the place of residency a less decisive factor for reproducing poverty, as is affirmed by Haroldo Torres (Torres, 2004).

Chart 8. Distribution of monthly family income by household characteristics (in R$)

	Family Income
Sanitary drainage – public network	333.90
Sanitary drainage – other	297.00
Water supply – public network	327.50
Water supply – other	261.30

Source: Federal Government Unified Record.

The relation between the destination of sanitary drainage and the monthly family income is notable. The increase given to family income, in this case, by the service offered by the Government, is the existence of a public network of sanitary drainage. Households situated in such places tend to have an average family income 12% higher than those who dispose of their waste by other means, such as rudimentary cesspools, septic cesspools, open air, ditches or other. The comparison between income and water supply was also very suggestive. Analogously, in places where there is public water supply network, the average family income tends to be higher (by 25%) than in households that collect water by other means (such as from wells, springs or water trucks).

The number analysis suggests that living in places devoid of certain fundamental public goods, such as a public water and sewage network, becomes an important element to determine the degree of family vulnerability and poverty. Therefore, encouraging public policies that promote "place gains" among beneficiaries, such as the increased coverage of public water and sewer networks, are fundamental to revert the aforementioned pernicious place effects.

To conclude, it is important to consider whether the income transfer is sufficient to reduce the poverty and reduce social inequalities.

Lena Lavinas (Lavinas, 2006) suggests that cash transfer programs should not be seen as panacea to resolve Brazil's social problems. In effect, we are not about to deny the positive effects of implementing such policies to reduce poverty; contrarily, we would like to show that they are insufficient to revert the inequality present in Brazilian society. The reason for this is relatively simple. We saw that among the factors that contribute most towards families leaving a situation of economic vulnerability is the possibility of accessing public services of a universalizing nature, such as health, education, sanitation, etc. As previously mentioned, daycare, having a retired person/pensioner in the family and having access to piped water, for example, are fundamental factors to increase average family income. We should not forget that these characteristics are universal rights, such as education from a certain age, retirement, sanitation, among others. Now, for this to happen, it is necessary that the Government offer these services.

When facing problems of such magnitude, it is noticeable that the cash transfer is important but not sufficient to overcome poverty and inequality. It is necessary that the Government promotes progress in universalizing policies, thus combining them with the income transfer ones. Attempts have been made, including within the previously mentioned "Brazil with no Misery" plan. It is important to mention that by doing so, the Government is actually putting into practice one of the fundamental dimensions that supports the "Bolsa Família" Program.

References

BOURDIEU, P. *The Misery of the World.* Petrópolis: Editora Vozes, 2003.

CORSEUIL, C. H.; FOGUEL, M. *Uma sugestão de deflatores para rendas obtidas a partir de algumas pesquisas domiciliares do IBGE.* Rio de Janeiro: Texto para Discussão IPEA, n.897, 2002.

LAVINAS, L. Transferências de renda: o “quase tudo” do sistema de proteção social brasileiro. *Anais do XXXIV Encontro Nacional de Economia*, 2006.

LÉVI-STRAUSS, C. Race and History. In: ______. *Structural Anthropology Two*. Rio de Janeiro: Tempo Brasileiro, 1987. p.328-66.

TORRES, H. D. *Políticas sociais e território*: uma abordagem metropolitana. São Paulo em Perspectiva, 18 (4), 2004.

12

An Indispensable Architectural Experience: João Filgueiras Lima – Lelé

Anália Maria Marinho de Carvalho Amorim[1]

Abstract

The class presented at the Top China Santander Universities, in July, 2010, the experience of one of the most important architect in Brazil, the Architect João Filgueiras Lima – Lelé, and intended to warn the youngster students not to give the production of their cities to repeatable processes of production of forms without the study of their consequences.

In July, 2010, at the *Top China Santander Universities* program, my chosen theme to be presented at Beijing University, to the Chinese and Brazilian students from different areas of interest and training, was an issue of interest to both countries: an experience that unites the concept of architecture and urban space in terms of the constructive logic of these cities.

The Program's focus for that year was environmental sustainability and urban planning, which reflects the need to rethink the planning strategies

1 Faculty of Architecture and Urbanism of University of São Paulo and Escola da Cidade – Faculty of Architecture and Urbanism, São Paulo, SP, Brazil. E-mail: *aamorim2000@dialdata.com.br*.

of cities around the world, and when conceiving urban growth, taking parameters of economic, cultural, social and environmental sustainability into account, without losing the characteristics of local realities.

Based on the fact that nothing compares to the process of urbanization city building that China has been going through during the past thirty years, I selected the work of Architect João Filgueiras Lima, also known as Lelé, so that the students could absorb the experiences of city creation as proposed by this architect.

The exhibition began by showing the creation of a simple park bench, from the first sketches to its execution. All the photos, designs and drafts presented here were provided by Architect João Filgueiras Lima.

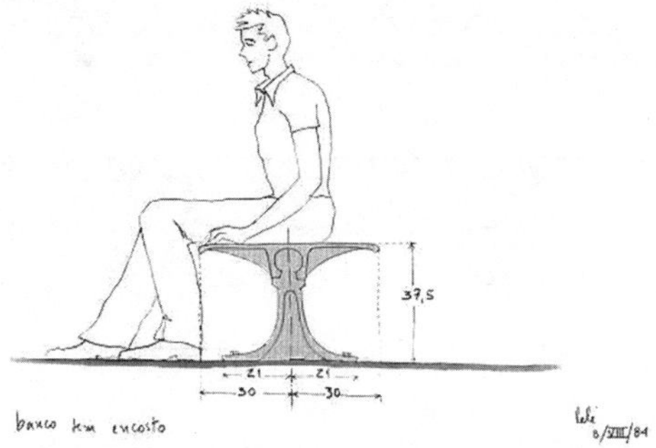

Drawings to conceive an object are governed by a structural logic and impositions of the constructive process.

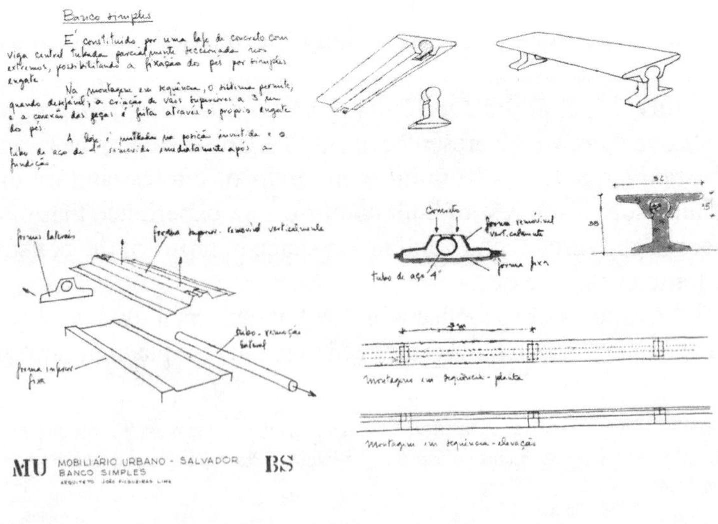

It was necessary to briefly describe that reinforced mortar as a constructive and structural material. Therefore, we displayed the steps of production for the bench, ranging from the production of the mold – something that can only be justified because the bench's construction will be repeated hundreds of times...

... to fitting the metallic screen that structures the bench in the mold…

... to the suspension and conduction of the mold, so that it can be filled by mortar…

...the filling of the mold...

... the placing of the molds in the pools for curing...

... the withdrawal of the bench from the mold...

... the mounting of variations of the same bench...

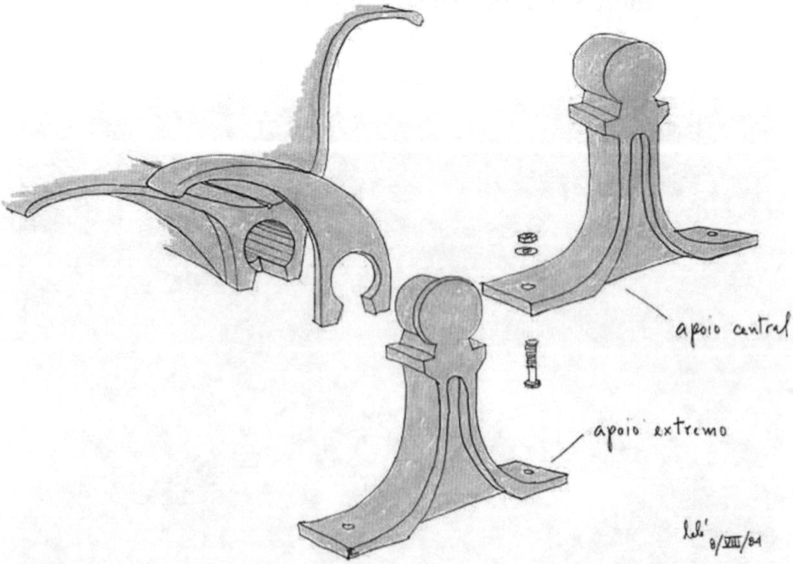

... the use of the bench in public areas...

The class aimed to demonstrate how the population of a country may be inserted into the production process of cities, because the exhibited architectural and construction equation counts on a fair amount of labor, which is within an intelligent system for production and assembly of equipment, infrastructure and housing.

Examples, using the experience of the same architect, exhibited the conception and construction of equipment that employs not only reinforced mortar, as structural and sealing elements, but also the logic of metallic structures and sealing.

It, The Sarah Hospital of Rio de Janeiro was shown as example, which as made entirely in the Technology Center of the Sarah Network, in Salvador, in the first decade of this century.

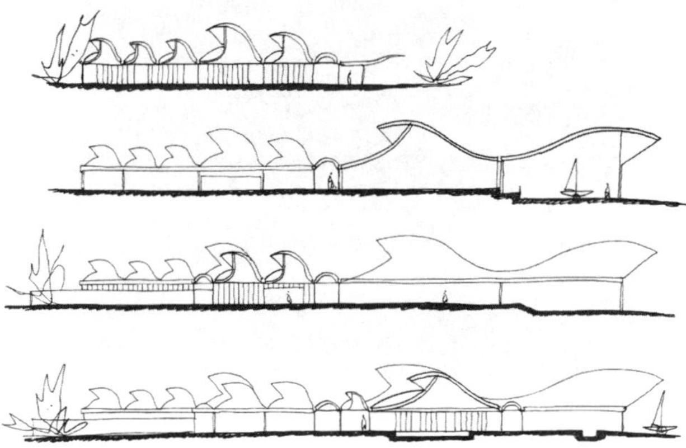

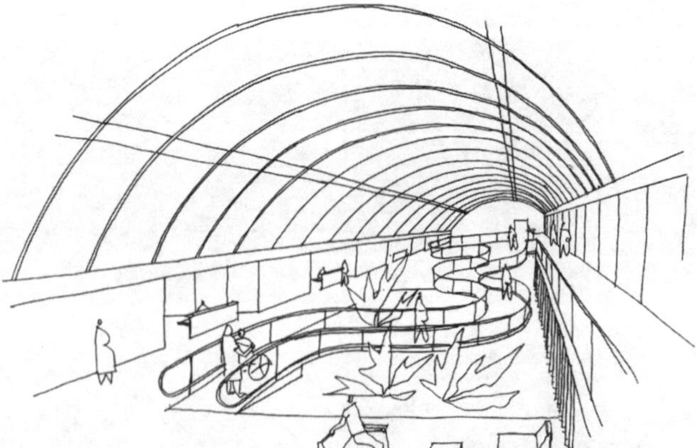

In addition to the class outlined above, my stay in Beijing with the Top China Santander Universities Program also resulted in a report to the Vice Rector for International Relations at the University of São Paulo.

The purpose of the report was to improve relations between USP and the Chinese universities, specifically between FAU-USP – Faculty of Architecture and Urbanism of the University of São Paulo and the Faculties of Architecture, Urbanism, Landscaping and Urban Planning in China in order to establish academic agreements with these institutions.

I spoke about potential study subjects that could unite us, issues that concern the two countries and also represent items of vital importance to the concerns of the planet.

For clarification purposes, I selected issues that have already been developed within the University of São Paulo by the Faculty of Architecture and Urbanism and issues that were highlighted during the trip to Beijing in July 2010 through the Top China Program.

The main items of research to be studied jointly by China-Brazil can be grouped according to the following aspects:

1. INFRASTRUCTURE:

- Environmental sanitation:
  Solid waste management
  Water resources management: procurement, storage and distribution, and water treatment; occupation of water source areas
- Urban Mobility and Accessibility
  Public Transport:
  Roads.
  Fluvial.
  Rail: China plans to build another 30,000 km of railways over the next 30 years. Within five years the entire country will have high-speed trains.

2. HOUSING

Scale versus quality.
Search and evaluation of construction materials
New cities: relation urban context and rural context
Cost of housing

3. URBAN DESIGN

Landscape Projects on different scales
System of Free Urban Spaces
Urban Parks
Design of Land-Water Borders

4. CONSTRUCTION OF URBAN FORESTS

Planning for Forest Construction and revegetation of urban centers

5. POPULATION DISTRIBUTION IN THE TERRITORY AND CREATION OF NEW CITIES IN FORMER RURAL AREAS.

Chinese Scenario: During the process of rapid urbanization, a large number of migrant workers came to town and became an important component of the city. The migration of people to the major centers frightens the government. Small towns in the country have 10 million inhabitants. There are more than 700 million Chinese in the country.

Social Housing project have been constructed in the country. There is a development program for small – and medium-sized rural towns. Currently there is greater fiscal and technical support for agriculture.

6. LAND POLICIES FOR THE COUNTRY AND CITY AND TERRITORIAL PLANNING

   Chinese Scenario: Management of the land confiscated by the state in 1949 and sold to be occupied by industry, housing...

7. URBAN FACILITIES

   Urban public spaces, such as open spaces for people to perform social communication activities , to meet and balance the needs of users.

8. ENVIRONMENTAL PLANNING

   Chinese Scenario: Change of energy production through the existing coal-fired thermal power stations, which represents 70% of the energy produced in China.
   Adoption of bio-fuels: second generation ethanol, alcohol from sweet potato and cassava.
   China already has the largest wind and solar farm in the world, but only 4% of its energy comes from it. The plan is that in ten years this percentage will be 15%.
   The country has exhausted its hydro-electric potential and has invested in the construction of five nuclear power plants.

9. REDUCTION OF HEAVY INDUSTRIES AND ADOPTION OF A MORE TECHNOLOGY-BASED INDUSTRY
10. SUSTAINABLE TECHNOLOGY AND CLIMATE CHANGE

11. STUDY OF CHINESE NATIONAL DEVELOPMENT PLANNING – National Develop Commission and its interfaces with academia, industry and environmental organizations.

12. ENVIRONMENTAL DEVASTATION OF THE COUNTRY AND POLLUTION OF WATER RESOURCES

Important issues discussed during the lectures and trip, in July 2010, at the Top China Program:

- INTERNAL ZONING OF AGRICULTURE AND UTILIZATION OF CHINESE FERTILE LAND

- EXPANSION OF ARABLE FIELDS TO OTHER LANDS. THE CASE OF BRAZIL AND AFRICA

- CONSUMPTION OF MINERALS. China annually consumes 650 million tons of steel, which is mainly used in civil construction.

This report has leveraged a process of opening agreements which have been consolidated over the last three years and has developed into several initiatives and research projects.

São Paulo, May 7th, 2013.

13

Sustainable Energy Sources in the Global Warming Scenario

Adir Janete Godoy dos Santos[1]

Introduction

The symptoms of a global decline in energy resources and biodiversity are clear, as are the increase in pollution and social and climate requirements. Global warming resulting from human interference, despite uncertainty about its extent, is now a fact that has been accepted by the scientific community (IPCC, 2007). Mankind should change its habits and adapt itself to use renewable energy sources, a process in which biomass and biodiversity will play a fundamental role. Renewable resources are less intensive than oil and mankind will not be able to substitute petroleum indefinitely, but it may support a society with lower consumption demands (Odum, 2001).

Climate change is one of the greatest challenges of the 21st century. Its most severe impacts may still be avoided if efforts are made to transform current energy systems. Data from the literature (IPCC, 2007; MMA, 2005; MMA-ANA-PNUMA, 2007) indicate that global warming has been caused by anthropogenic action. According to these reports, the main source of historic and current emissions of

1 Anhembi Morumbi University, São Paulo-SP, Brazil. E-mail: *godoyaj@hotmail.com*.

CO_2 is from the burning of fossil fuels. Since 80 per cent of global carbon dioxide emissions derive from combustion of fossil fuels, it is clear that radical changes in production and consumption patterns are necessary, including resources with an energy evaluation matrix (MCT, 2009), cleaner production procedures, and impact assessment of solid residues and liquid effluents generation, which are released to environment (Santos, 2003; Santos, 2005; Nisti; Santos, 2008). These practices contributes to increasing of global warm.

Access to services such as energy, transportation, telecommunications and water significantly improves the human development index (IDH) and also lays the ground for continuous economic growth. These are critical inputs for the provision of goods and services and significantly affect economic development productivity, cost, and competitiveness. One important theme regarding the current global agenda is conciliated with a high IDH, natural resource preservation and cleaner energy sources being inserted into human production. Its importance derives not only from economic reasons, but also from the need to assess emissions of green house gases into the atmosphere, as well as all other kinds of gaseous emissions, solid residue, waste and liquid effluents released into the sanitary system and environment.

Any anthropogenic activity must consider an environmental impact assessment inherent to productive practices in the industrial field, including alterations in natural resources: water, terrestrial ecosystems and the atmosphere (Santos, 2003). A continuous and representative effluent monitoring program was established at the production plant in order to allow the environmental impact assessment, which is an efficient tool for systemic management when faced with environmental regulatory rules. (Nisti; Santos, 2008).

The generation and storage of industrial solid waste requires control, monitoring and impact assessment for its occupational exposure and public increment impact (Santos, 2003; Santos et al., 2006). The chemical types present in the several phases of solid waste are disposable for environment or human, depending on the composition, interstitial water, chemical and physical characterization (Santos, 2005; Santos; Mazzilli; Favaro, 2006) and the need to evaluate workers' exposure, using occupational parameters (Santos et al., 2006).

The consolidating results in effluents' systemic monitoring generates a statistical database and contributes towards cleaner production (Santos; Nisti, 2011) and the development of cleaner mechanisms.

Energy Sources in Brazil

The current price of oil and the questions regarding global warming associated with the greenhouse effect have led Brazil to reconsider increasing its share of renewable energy in its energy matrix over the next few decades. Brazil must keep its ethanol program and has already announced their intention to increase the number of ethanol and hydroelectric plants (MME, 2007).

Since climate change is a global challenge, international action is essential. In Brazil, concerted action must be centered on the international regime on climate change embodied in the United Nations Organization Framework Conventions, the Kyoto Protocol and international agreement based on these (DOU, 2008).

The current Brazilian generation matrix is one of the cleanest in the world. The participation percentage of non-renewable sources was about 55% for 2006, 2007 and 2009. The contribution of renewable sources was about 45% during same period. Biofuels (sugar cane ethanol, charcoal and biodiesel), obtained by biomass transformation, contributed with about 30% in the energy matrix (BEN, 2008, 2010; Santos, 2008). Renewable energy sources have a great potential to displace greenhouse gas emissions from the fossil fuel combustion and thereby mitigate climate change.

Table 1 shows the Brazilian energy matrix, with contributions in percent (%), for each fuel, from renewable and non-renewable, in 2006, 2007 and 2009. The data (%) were compared with that of countries in the Organization for Economic Co-operation and Development (OECD) and with the global percentage (Tomasin, 2010).

Table 1. Brazilian energy matrix, fuel resource contribution (%) in 2007 and 2006 (BEN, 2008, 2010; Santos, 2008, Tomasin, 2010)

Energy Source	Energy Resource	Brazil (%)			World (%)	OECD
		2006	2007	2009	2007	2008
Non-renewable	Oil and derivative	37.8	36.7	37.8		
	Natural gas	9.7	9.3	8.9		
	Coal and derivatives	6.0	6.2	4.8		
	Nuclear – (U_30_8)	1.6	1.4	1.2		
	Total	55.1	53.6	52.7	87.3	92.8

Energy Source	Energy Resource	Brazil (%)			World (%)	OECD
		2006	2007	2009	2007	2008
Renewable	Sugar Cane (ethanol)	14.5	16.0	18.1		
	Hydraulic and Electric	14.8	14.7	15.3		
	Charcoal and firewood	12.7	12.5	11.8		
	Biodiesel and others	2.9	3.1	2.1		
	Total	44.9	46.3	47,3	12.7	7.2

As can be seen in Table 1, the total percent share for renewable energy sources in Brazil is very significant at around 45%, when compared with the international statistical average, which is 12.7 % and 7.2 % of renewable sources, in the world and the OECD Members States, respectively.

In Brazil, from the non-renewable fossil fuels, mineral coal can be found with abundant reserves and large reserve/production and reserve/consumption ratios, excluding metallurgical coal. Brazil is almost completely dependent on imports of these commodities of metallurgical grade (about 11 million equivalent metric tonnes oil in 2007) (DNPM-MME, 2007; Santos, 2008).

When only electricity production is considered, Brazil could be called mono-energetic. The main field of electricity production is hydroelectric, which participates with about 85% in 2009 and over the last few decades (Ben, 2010; Tomasin, 2010).

As is presented in Table 2, the renewable source share (%) is about 90% and 18%, in the Brazil and in the World, respectively.

Table 2. Renewable contribution (%) in electricity share in Brazil and in the World (BEN, 2010; Tomasim, 2010)

Energy Source	Eletricity	Brazil (%) 2009	World (%) 2007
Non-renewable	Oil and derivative	2.9	5.6
	Natural gas	2.7	20.9
	Coal and derivatives	1.3	41.5
	Nuclear – (U308)	2.5	13.8
	Total	9.4	81.8

Energy Source	Eletricity	Brazil (%) 2009	World (%) 2007
Renewable	Hydraulic and Electricity	85.0	15.6
	Biomass (*)	5.4	–
	Others	0.2	2.6
	Total	90.6	18.2

Over the last 50 years, Brazil has diversified its production and use of energy. The planning encompassed various objectives, including diminishing the risk of energy shortages, energy efficiency, supplying demand at relatively low costs and controlling environmental and social impacts. Renewable sources of energy may be essential for Brazil to achieve the commitments set out in the parameters of these goals, which must take environmental benefits into consideration, such as reducing greenhouse gas emissions, technological and scientific development, social development and energy cost reduction that can result from the use of renewable energy sources. As is presented in Table3, in Brazil, the greatest renewable energy share was in the 1970-1980 decade.

Table 3. Energetic source participation (%) from decades 1970 to 2007 (FIESP, 2001; Santos, 2008)

Year	Brazil Source Participation (%)	
	Non-renewable	Renewable
2007	53.6	46,4
2000	59.0	41.0
1990	50.9	40.19
1980	54.4	54.6
1970	41.6	58.4

The development of ethanol production in Brazil started in the 1970s, which was based on public policies adopted under the Programme of Incentives for Alcohol Production (Proalcool), with the aim of diversifying this fuel source to vehicle transportation in the country by increasing the use of new alternative energy sources.

After 1970, the big plants option (120 m^3/day or more) was chosen despite the other possibilities that existed at that time. As a result, there was a great destruction of natural and human resources over the last decades such as, native forest destruction, rural exodus, and diversified farm land occupied by sugar cane plantations. The preference for great scale resulted in monoculture, loss of soil fertility, land ownership concentration and biodiversity loss. This choice also affected the quality of the global atmosphere, due to CO_2 emissions that come from indirect use of fossil fuels.

The great advantage of biofuel ethanol was the direct increase in CO_2 emissions of into atmosphere, when we consider the global biogeochemical cycle of carbon and biomass sugar cane growth. In this cycle mechanism, the carbon in labile pool was dislocated with a higher velocity, with it being accumulated in the vegetal specie.

The main disadvantage of Pro-alcohol did not consider the possibility of organic biodegradable residue being used, which is a potential biomass and possible raw material in the process to obtain ethanol.

One FIESP-CIESP (2001) study recommends that all industries located in São Paulo State improve their efforts in bioenergy production by using organic industrial waste, which demonstrates the viability for costs, environmental legacy and mitigation of impacts in the soil, water, biodiversity and atmosphere.

According to Electrical Energy Ten-year Plan 2017 (MME, 2007a) the estimated demand for ethanol domestic consumption will increase from about 20 to 50 billion liters between 2008 and 2017, respectively. The calculation was obtained by adopting the internal usage in domestic transportation (Santos, 2008; Santos, 2009). The data are presented in Figure 1. The current impacts require urgent assessment.

Figure 1. Sugar cane ethanol demand in billion liters per year, 2008 to 2017.

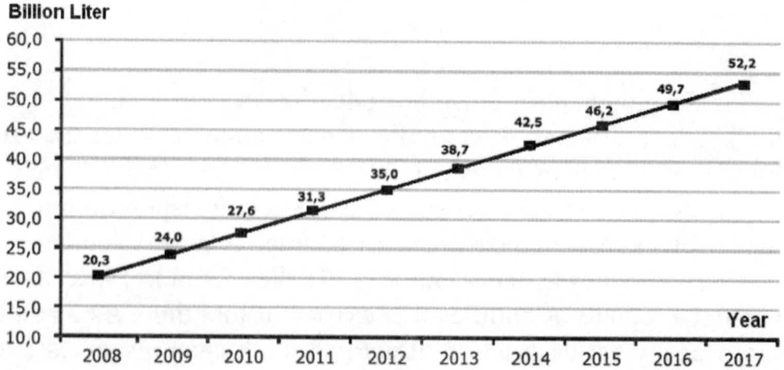

Planning and Regulation

Energy planning was practically abandoned in the 1990s and evaluated after 2002 with basic premises of growth below that what the government desired. The 2030 Plan assumed that the gross domestic product (GDP) would grow by 4% per year (2005-2025), or an equivalent growth of 3.2% GDP per capita (MME, 2007).

Today, the process of energy planning is dynamic and enables new situations and improvements to be incorporated in the expansion of criteria and investment. The new approach for energy source generation should be outlined according to the technological capacity and potential.

The objectives for Brazilian energy matrix expansion must be in compliance with the Ten-Year Energy Plan – PDE, (MME, 2007a), National Energy Plan 2030 (MME, 2007). This could help provide the long-term required for governmental and energy industry actions, which is so it can develop into a more secure and stable environment, to obtain a focus on sustainability in terms of environmental, social, economic, cultural and territorial requirements.

A joining of the environmental legal framework and international agreement would provide certainty in terms of the development volumes in the medium- and long-term, legal security in the processing of projects, and a support system that would enhance competitiveness.

The Ten-Year Energy Plans in Brazil were consecutively modified each year between 2006 and 2013 (it is under public review today). The contention surrounding the monetary spending by the government has been not adopted in these studies, which rapidly changing conclusions and goals changed and were the most financial expensive planning scheme for the government and the population.

For example, the 2010-2019 Electrical Energy Ten-year Plan contains annual expansion forecasts to generate wind energy. This source energy choice is recent. The development of wind power in Brazil started in 2002 and was based on the public policies established in the Incentives for Alternative Electricity Sources Programe (PROINFA), the aim of which being to diversify from the electricity share in the country by increasing the use of this energy source Future requirement for wind power industry goals need the establishment of a legal framework in this sector.

The requirements for energy matrix planning are presented in Figure 2.

Figure 2. Transversal diagram of the Brazilian energy matrix improvement (modified by Santos, 2008)

The main discussion in the National Energy Plan – PNE 2030 (MME, 2007) seems to be related to the demand for natural gas. This demand would reach 26 billion m^3 per year in that year, corresponding to approximately 6% of current Brazilian reserves, which includes the inferred reserves. Electricity generation is one of the uses for natural gas, accounting for approximately 9% of the energy matrix in Brazil, as is presented in Table 1. Consumption of natural gas in 2030 would correspond to almost 30% of current reserves in only one year.

Hydroelectric energy is still advantageous in terms of generation costs. However, environmental deadlocks prevent its potential from being fully utilized. On the other hand, future demand will be concentrated in the southeast and northeast of the country, which is where the hydroelectric potential would soon be exhausted if it were fully exploited (Alvim, Eidelman and Mafra, 2007).

The largest power plants are being planned for the Amazon region, where environmental difficulties are even greater. But it seems probable that, because of its economic advantages, considerable hydroelectric generation capacity will be installed in that region.

Decisions and Goals

Energy demand under sustainability (economic, social, environmental, cultural and territorial) in current Brazilian planning are complex and requires the integration of different fields, as is presented in Figure 2.

The comparative environmental, technological and strategic costs, including national reserves, consumption demand and the non-proliferation international agreement must construct a comprehensive nature over the medium- and long-term.

Brazilian biofuels, such as ethanol and biodiesel, are uncontestable sources of wealth for the country. Their use replaces fossil sources that have a great impact on the climate and on air quality. The matrix needs to consider the monoculture questions and deforestation impacts.

Organic solid waste, such as raw material for biofuel energy, provides a comprehensive response to the principles of sustainable and competitive development, which is implicit in the 1988 Constitution of the Federative Republic of Brazil (DOU, 1988), and contributes to an efficient implantation of the National Solid Residue Politic (MMA, 2010), thereby justifying a very substantial increase in its share in the energy grid.

The reference scenario in the National Energy Plan – PNE 2030 (MME, 2007) stipulates the expansion of the electricity supply between 2005 and 2030 will include 95,000 MW from hydropower plants.

The growing demand and the prospect for the long-term exhaustion of hydroelectric potential, in addition to questions of a socio-environmental nature, require diversified sources as part of this expansion, despite the priority that hydropower deserves in the expansion of the Brazilian electrical energy system.

A large grand part of the hydropower plant is just in the operation or building. The building of new plant in the current evaluation requires high financial investment and a long construction time.

Acknowledgements

The author would like to thank to Santander Bank for their successful planning, organization, financial support, and incentives in the Santander Universities Top China Programme – Top China 2009.

Also to Global Division Santander Universities for promoting and stimulating different areas of development and for supporting the transfer of knowledge from academia to society, which includes the professors and students from around the world. The author is grateful to Shangai Jiao Tong University for integrating the programe, to Universidade Anhembi Morumbi-UAM for their technical and scientific effort, to Universidade de São Paulo – USP for their collaboration and information in the Top China 2009 organization, to Universidade do Estado de São Paulo – UNESP for their collaboration in this paper.

References

ALVIM, C. F. et al. Nuclear energy over a 30-year scenario. *Estudos Avançados*, 21 (59), p.197-219, 2007.

BEN. *Balanço Energético Nacional 2008*. Brasília: Empresa de Pesquisa Energética – Ministério de Minas e Energia, 2008.

DNPM-MME. *Sumário Mineral*. Brasília: Departamento Nacional de Produção Mineral – Ministério de Minas e Energia, 2007.

DOU. *Constitution of the Federative Republic of Brazil*. Diário Oficial da União, Brasília, 1988.

FIESP. *Ampliação da oferta de energia através da biomassa*. São Paulo: Federação das Indústrias do Estado de São Paulo, 2001.

IBAM. *Mecanismo de desenvolvimento limpo aplicado a resíduos sólidos*. Rio de Janeiro: Instituto Brasileiro de Administração Municipal, 2007.

IPCC. *Mudança Climática 2007*: Impactos, Adaptações e Vulnerabilidade. Contribuição do Grupo de Trabalho II. Painel Intergovernamental Sobre Mudança Climática. PARRY, M. L. et al. (Eds.). Cambridge, Reino Unido, Cambridge University Press, 2007.

MCT. *O mecanismo de desenvolvimento limpo guia de orientação*. Rio de Janeiro: Ministério de Ciência e Tecnologia; Imperial Novo Milênio Editora, 2009.

MMA. *Interministerial Committee on Climate Change* – Decree N.6263 of November 21th 2007. Sumário Executivo Brasil Plano Nacional Sobre Mudança Climática. Brasília: Ministério do Meio Ambiente, 2008.

MMA-ANA-PNUMA. *GEO Brasil* – Recursos hídricos resumo executivo. Brasília: Ministério do Meio Ambiente. Agência Nacional de Águas. Programa das Nações Unidas para o Meio Ambiente, 2007.

MMA. *Instrumentos econômicos para o desenvolvimento sustentável na Amazônia brasileira*. MAY, P. H. et al. (orgs.). Brasília: Ministério do Meio Ambiente, 2005. p.124.

MME. *Plano Nacional de Energia* – PNE 2030. Brasília: Ministério de Minas e Energia – Empresa de Pesquisa Energética, 2007. 408p.

MME. *Plano Decenal de Expansão Energética* PDE 2017. Brasília: Ministério de Minas e Energia – Empresa de Pesquisa Energética, 2007a.

MMA. *National Solid Residue Politic*. Brasília: Ministério do Meio Ambiente, 2010.

NISTI, M. B.; SANTOS, A. J. G. *Assessment of radioactive liquid effluents release at IPEN-CNEN/SP*. The Natural Radiation Environment: 8th International Symposium (NRE VIII). American Institute of Physics – AIP Conference Proceedings, v.1034, p.505-8, 2008.

ODUM, E. *Ecologia*. Rio de Janeiro: Editora Guanabara Dois, 2001.

ONU. *Cúpula Mundial Sobre Desenvolvimento Sustentável*. United Nations Organization. Jhoanesburg, South Afric, August 26-September 4, 2002.

SANTOS, A. J. G. 2002. *Radiological environmental impact assessment of the Brazilian phosphogypsum and leaching of* ^{226}Ra *e* ^{210}Pb. PhD Thesis, Universidade de São Paulo, São Paulo, 165p. In Portuguese.

SANTOS, A. J. G. *Regulamentação e decisões na avaliação de impacto ambiental dos resíduos contendo radioatividade*. International Nuclear Atlantic Conference INAC. Santos, Brazil, 2005. ISBN: 85-99141-01-5.

SANTOS, A. J. G.; MAZZILLI, B. P.; FAVARO, D. I. T. Partitioning of radionuclides and trace elements in phosphogypsum and its source materials based on sequential extraction methods. *Journal of Environmental Radioactivity*, v.87, p.52-61, 2006. ISSN:0265-931X.

SANTOS, A. J. G. et al. Radiological characterization of disposed phosphogypsum in Brazil: evaluation of the occupational exposure and environmental impact. *Radiation Protection Dosimetry*, v.121, n.2, p.179-85, 2006. ISSN: 0144-8420.

SANTOS, A. J. G. *Mesa Redonda*: Alternativas Para a Produção de Energia Limpa. II Mostra Sistema FIESP de Responsabilidade Socioambiental. São Paulo, 13-15 agosto 2008.

SANTOS, A. J. G. *Energetic sources and the global warming.* Santander Universities Top China Programe – Top China 2009. Shanghai, July 25 to August 15, 2009.

SANTOS, A. J. G.; NISTI, M. B. *Decisions and procedures to cleaner production concerning on liquid effluents assessment.* 3rd International Workshop Advances in Cleaner Production. p.1-11, São Paulo, May 18-20, 2011.

TOLMASQUIM, M. T. *The energy market in Brazil.* Seminar el desafío energético del Siglo XXI. La estrategia de los países vecinos frente a la crisis energética. p.1-8. 2010.

14

Better use of scientific knowledge for green, clean and sustainable cities

Luciana Pinto Sartori[1]

Introduction

Environmental issues are a common concern, and countries with huge areas and population are facing situations where development can share space with nature. Well-planned cities count on the advances of architecture to generate green and sustainable areas. People are valuing spending more time in parks and open areas, trying to avoid polluting environments more, and exploring the advantages of having a clean river to navigate and even enjoying beautiful parts of the city (Photo 1).

Many countries were even able to reduce pollution and touristically exploit their main rivers, such as the Seine in Paris and the Thames in London (Andrews, 1984). It would be a great achievement if a city like São Paulo could reach this level of organization and success regarding the cleaning of its water, and therefore, the entire city.

The importance of open spaces for public use, which are free of buildings and are rich in vegetation, in cities is associated with quality

1 Biological Sciences Professor, Centro Universitário São Camilo, São Paulo, SP, Brazil. E-mail: *lpsartori@hotmail.com*.

of life and must be considered during cities' planning and organization. These are essential for ecological and socio-economic purposes, allowing the appropriate functioning of natural systems, which purify water; harbors plants and animals; clean the air and regulate the climate.

Photo 1. A night cruise on the Huangpu River, Shanghai.

It is extremely important to call attention to careful planning as a way of organizing the urban territory, and as a means of determining interventions for rational occupation, recommending leisure in underused spaces (Ribeiro; Silveira, 2006).

In Brazil, one city that is most famous city for its ecological approach is the city of Curitiba, which is the capital city of the State of Paraná. In Brazil there are many capitals that deal with damage from summer floods. Most green areas in Curitiba were planned in accordance with the need to contain and restrain excessive rain, which even includes artificial lakes that were inserted into recent planning so that water accumulation could be regulated. Other parks were planned to avoid pollution close to bodies of water, which was done in such a way so as to have less houses surrounding the area close to the riparian forest, which are protected by law. It also values the area, so, in terms of rental state business, it is also a big advantage to live close to parks, including better air quality, since the presence of the trees generates better quality and colder air.

The creation of parks in cities is not only focused on providing citizens with a better quality of life, but also on generating profit for the city as a whole (Buccheri Filho, 2012). Parks can be considered as an important vehicle to develop the municipal district, within a sustainability concept (Ribeiro; Silveira, 2006).

Especially in big cities, where distance is an issue, it is fundamental for the families to find open areas to spend time while being integrated with the environment. There has been a visible growth in public spaces being used by inhabitants and tourists for sport, walking and biking. Adjustments regarding people with reduced mobility must also be considered in the projects.

Importance of Parks

Parks offer the opportunity to spend time outdoors, therefore leisure promotes growth of towns, improvement in quality of life and generation of jobs through new sectors such as ecotourism. All of which depend on well-planned urban infrastructure, with areas designated for consumption. By improving the quality and quantity of open public open spaces it is possible to make urban life more interesting and meaningful through the creation of livable environments (Tureal; Yigit; Altug, 2007).

A very good example is the Botanical Garden in Curitiba, which was previously a disused space, but, following planning, the urban landscape was altered which stimulated tourism and environmental education (Ribeiro; Silveira, 2006).

Besides botanical gardens there are public parks, zoos and community orchards that offer inhabitants and tourists the opportunity to enjoy outdoors activities, all of which depending on good planning and management. Museums and galleries are some of the most visited touristic attractions in a big city, but open areas attract a different kind of tourist, and are also an important place for elderly people (Farquhar, 1995; Tureal; Yigit; Altug, 2007).

In Table 1 it is possible to observe the percentage use of open areas in the city of Curitiba. These areas are still used less than museums and malls, but this is increasing according to data from the Secretary of Tourism of Paraná. Using this kind of information can incentivize other cities to better explore its open areas for sightseeing uses.

Table 1. Some of the touristic attractions in Curitiba (State of Paraná, Brazil).

Touristic attraction	(%)
Parks	22,2
Botanical Garden	17,5
Zoo	3,4
Public Garden	3,4
Green areas	1,6
Sightseeing Bus	4,2

Source: Paraná Turismo, 2013.

In organized cities such as London, where tourism is well planned and valued, botanical gardens and parks can almost account for 50% of the attractions demanded by tourists (Table 2).

Table 2. Percentage of the touristic attractions in London (England).

Touristic attraction - London	(%)
Parks and wild life sites	11
Botanical Garden	29

Source: Insights UK, 2013.

Elderly people use park areas mostly as a place to meet friends and socialize, which is part of their needs to reach a good quality in life, since social contact is as valued a component of a good quality of life as good health (Farquhar, 1995). Therefore, when urban green spaces are available, there are wider nonphysical benefits such as impacts on wellbeing and mental health, as well as social inclusion (Lee; Maheswaran, 2010).

In China people live longer than in many other countries, so they are considered in programs designed for parks and gardens that give them discounts or free passes, since they are frequent users and have much more free time than other users. This is the case for Shanghai Botanical Garden, the largest municipal botanical garden in China which is also a garden that develops research with plants and has 500-year- old bonsais. It is more than a touristic attraction, it is also part of the Chinese culture, since people enjoy its secular beauty and also use it for Yoga, Tai Chi (Photo 2) and meditation.

Photo 2. Tai Chi in an open area at the Summer Palace (Beijing).

Green Building

Eco-friendly architecture searches for the concept of environmental health as well as that of human well-being. It is still a distant dream for some people to have an ecological house, but when it becomes popular, the price of transforming old houses into eco-friendly ones will be accessible to most people. All projects related to green buildings focus on increasing the efficiency in terms water and energy use, which decreases common monthly expenses. Thus, according to the principles of sustainable construction, it is possible to reduce resource consumption; reuse resources; use recyclable resources; protect nature; eliminate pollution; apply life-cycle costing, all of which focus on quality, mostly quality of life (Kibert, 2012).

Urban structuring can combine modern roads with beautiful gardens, which brings all sorts of benefits since people enjoy taking the sightseeing bus to observe the combination of gardening and art (Photos 3 and 4).

Photos 3 and 4. Shanghai combines green areas with modern constructions and art.

Environmental Issues

Biological invasion

Biological invasion is a serious environmental problem all over the world. Species introductions and invasions are due to human activities and threaten native ecosystems, mainly the biodiversity, due to the competition established by foreign species with the native ones (Yang; Nakagoshi, 2004).

There is a common coastal plant that is extremely important in Brazil named *Spartina alterniflora* that is considered an invasive species in many countries (Burfeind et al., 2013) and is causing a great deal of trouble to the Chinese who introduced it to their coast. In Brazil it is classified as an important shelter area for macrofauna associations (Flynn; Tararam; Wakabara, 1996), and is even used as a bioindicator for estuarine quality (Stein; Saiki; Braga, 2011). *Spartina* is one of the greatest invasive species listed by the State Environmental Protection Administration of China. Successful plantations and the rapid spread of this species have increased the production and fertility of the coast, but at the cost of native ecosystems (Chen et al., 2008). Its transplanting to Chongming East Beach in Shanghai has led to discussions since its intentional introduction, and now it can be found in almost all parts of China's east coast and has provoked a variety of responses, with it being even considered a pest (Yang; Nakagoshi, 2004).

This is a clear reflextion of poor use of knowledge regarding a very well-studied plant. This exposes the urgent need to apply academic studies within the city's planning and management. Chen et al. (2008) conclude that it is time to listen to the scientific community and take a new direction towards stopping further plantations of *Spartina*, and instead use native plants along coastal wetlands.

China (including Taiwan, Hong Kong and Macao since they are all considered a part of China) is one of the world's richest countries in terms of biodiversity and is extremely vulnerable to invasive species of foreign origin. It is a fact that legislation is needed to restrict use of alien species, and to encourage the use of domestic species, since non-native species are a major threat to biodiversity and must have a management plan with clear guidelines (Burfeind et al., 2013; Yan et al., 2001).

Endangered species

There are many endangered animal species in China that people know well, that are even the symbol of big groups (such as the Panda for WWF), but there are few animals that people do not know about, that can be seen and visited but are not well covered, in an informational sense, regarding their condition and dwindling population.

One example is the South China Tiger (*Panthera tigris amoyensis)* which is considered, by the IUCN (International Union for Conservation of Nature) as a critically endangered species, its population is already considered extinct in the wild. The first problem for this species was being considered a pest in the past, so now only few individuals remain in Zoos (such as in the Beijing Wildlife Park – Photo 5).

Some wildlife parks are successful in saving animals, as well as their protection and reproduction, but some also exploit the animals to a dangerous limit in order to be more profitable. This is the case when they allow people to hold baby tigers, and baby pandas, which are all endangered species and have very complicated reproduction development processes in captivity. Also, in the same context, playing with the snub-nosed monkey, since they already achieved a large artificial population, is considered by some to be educational, which involves exposing these animals to humans will. In the photo one can see that by just paying it is possible to sit with the animal and take a picture in the Beijing Wildlife Park (Photo 6). The golden monkey *Pygathrix roxellana* or *Rhinopithecus roxellana* is one of the most beautiful and rare animals in the world that is unique to China, the animal is listed as one of the world's endemic endangered species (Sharma, 2005).

When visiting the Great Wall of China it is possible to pay to take a photograph with the endemic wild Chinese Camel, also known as the Bactrian camel. But few people would pay attention to this species when visiting this amazing sightseeing location (Photo 7). Wild camels (*Camelus ferus*) are genetically distinct from the domesticated Bactrian camel and have smaller humps. They are considered to be the only land mammals capable of drinking salty or brackish water if there is no fresh water available. Since 2002 they have been classified as critically endangered. They are threatened by hybridization with domestic camels and wolf predation, and the fact that the genetic information of the wild two-humped camel remains obscure means that implementing conservation programs is difficult (Ji et al., 2009).

Photo 5. Chinese tigers can be seen in a large open area, visitors can take pictures from a bus.

Photo 6. The Golden or Snub-nosed monkey. (Beijing Wildlife Park, July, 2012)

金丝猴
在树上
果等为
6个月，
Rhinopithecus roxellanae
monkey
疣猴科
川、云南、
Golden monkey
Shanxi
5000
down
is leav
and bi
is 6 m
matur

Photo 7. The Chinese Camel (Bactrian) at the Great Wall of China (July, 2012)

The Giant Panda (*Ailuropoda melanoleuca)* is another animal considered to be an endangered species by the IUCN, and is the most powerful symbol for species conservation. In China, the Giant Panda is a national treasure, and, for WWF, the panda has a special significance since it is the symbol of their organization. These animals have special areas in the zoos and are kept in adjusted temperature environments, with good amount of food offered to them, in addition to having access to an area far from the visitors.

Responsible zoos and aquariums should strive to conserve ecosystems and promote care with positive action for the animals. Not only the endangered species should be taken good care of, but also those that are far from their environment and exposed to people. However, this seems to be far from real when it comes to exploiting both animals and visitors, since it is always possible to touch the animal by paying for it (Photo 8), but it only happens when there is no legislation or enforcement of regulations.

Photo 8. Giraffes in the Beijing Wildlife Park being given food by children.

Conclusions

It is time to bring together knowledge in terms of geography, planning and management so that the unused spaces in urban areas can contribute to a better quality of life (Cavalheiro, 1991). If cities as big as São Paulo and Shanghai could be directed with an environmental approach, exploring the knowledge taught at the universities, then fewer mistakes would be committed, better profits would be taken for both citizens and tourists.

It would be possible to avoid biological invasions, which cause high biodiversity loss and environmental damage. It would be possible to protect endangered species and use environmental education to create new generations of concerned citizens. It would be possible to have clean rivers as well explored as navigation systems, which are used as a touristic attraction and humidity controller.

Acknowledgements

The author would like to thank Santander Bank and the Assessoria de Extensão, at University São Camilo/SP.

References

ANDREWS, M. J. Thames Estuary: pollution and recovery. In: SHEEHAN, P. J. et al. (Eds.). *Effects of Pollutants at the Ecosystem Level*. New York: John Wiley and Sons, 1984. 227p.

BUCCHERI FILHO, A. T. O planejamento dos parques no município de Curitiba, PR: planejamento sistemático ou planejamento baseado em um modelo oportunista? *Caminhos de Geografia*, 13(41), p.206-22, 2012.

CAVALHEIRO, F. Urbanização e alterações ambientais. In: TAUK, S. M. *Análise ambiental*: uma visão multidisciplinar. São Paulo: Unesp-Fapesp, 1991. p.88-9.

CHEN, J. et al. Invasive Spartina and reduced sediments: Shanghai's dangerous silver bullet. J Plant Eco, 1(2), p.79-84, 2008.

FALK, J. H. et al. *Why Zoos & Aquariums Matter*: Assessing the Impact of a Visit. Silver Spring, MD: Association of Zoos & Aquariums, 2007.

FLYNN, M. N.; TARARAM, A. S.; WAKABARA, Y. Effects of habitat complexity on the structure of macrobenthic association in a *Spartina alterniflora* marsh. *Rev Bras. Oceanogr*, 44(1), p.9-21, 1996.

JI, R. et al. Monophyletic origin of domestic bactrian camel (*Camelus bactrianus*) and its evolutionary relationship with the extant wild camel (*Camelus bactrianus ferus*). *Animal Genetics*, 40, p.377-82, 2009.

KIBERT, C. J. *Sustainable Construction*: Green Building Design and Delivery. 3rd. John Wiley & Sons, 2012. 560p.

LEE, A. C. K.; MAHESWARAN, R. The health benefits of urban green spaces: a review of the evidence. *Journal of Public Health*, 33(2), p.212-22, 2010.

RIBEIRO, R. M.; SILVEIRA, M. A. T. Planejamento urbano, lazer e turismo: os parques públicos em Curitiba – PR. *Turismo, Visão e Ação*, 8 (2), p.309-21, 2006.

SHARMA, C. *Chinese Endangered Species at the Brink of Extinction*: A Critical Look at the Current Law and Policy in China Animal L. 215. 2005.

STEIN, C. E.; SAIKI, M.; BRAGA, E. S. *Spartina alterniflora* como macrofito indicadora de qualidade ambiental em estuários – caso do complexo estuarino-lagunar de Cananéia/Iguape. *Anais: V Simpósio Brasileiro de Oceanografia*. Santos, SP, 2011.

TUREL, H. S.; YIGIT, E. M; ALTUG, I. Evaluation of elderly people´s requirement in public open spaces: a case of study in Bornova District (İzmir, Turkey). *Building and Environment*, 42(5), p.2035-45, 2007.

YAN, X. et al. Invasive species in China – an overview. *Biodiversity and conservation*, 10, p.1317-41, 2001.

YANG, F.; NAKAGOSHI, N. Alien plant invasion in water systems in Shanghai, China. *Journal of International Development and Cooperation*, 10(2), p.1-11, 2004.

Websites

http://www.insights.org.uk/articleitem.aspx?title=Gardens+Tourism+-+is+the+Market+Really+Blooming%3F (Access May, 2013)

IUCN – http://www.iucnredlist.org/

WWF – http://wwf.panda.org/

15

Bacterial regrowth and its control in drinking water distribution system

Yiming Liao; Xiaohui Bai[1]

Introduction

At present, water distribution is facing a challenge regarding how the low concentration of leaking microbes results in large amounts of bacterial regrowth (Kistemann; Dangendorf; Exner, 2001; Kerneys et al., 1995). Bacterial regrowth bacteria affects water quality in general, not only concerning healthy risks, but also involving sensory problems and the corruption of distribution components (Safe Drinking Water Committee Biological quality of water in the distribution system[M], 1982). However, controlling microbial regrowth in a distribution system is a complex systematic problem, which include water source selection, raw water purification, pipeline distribution control and regulation in the secondary water supply (David; Mossel, 2004). As for water quality, the positive regulatory effect of nutrients such as organic carbon, nitrate and

1 Lab of Water Science and Technology, School of Life Sciences and Biotechnology, Shanghai Jiaotong University, Dongchuan Rd., Shanghai, 200240, China. E-mail: *xhbai@sjtu.edu.cn*.

phosphorus conflicted with the negative restriction of disinfection that may affect regrowth patterns altogether (Power; Nagy, 1999; LeChevallier et al., 1996).

The concept of biological stability was introduced in order to restrict microbial regrowth, which estimates the maximum possibility of microbial regrowth in the water by the means of nutrient substrates (Hammes et al., 2007). In Europe, biological stability has been a topic of great interest in order to ensure that microbial regrowth is restricted in the drinking water distribution system since the 1980s (LeChevallier, 1990, 82(11)), and it has become an acknowledged concept that is used in both drainage and distribution systems.

Biological stability is usually characterized by the following parameters: AOC, BDOC, BRP, MAP etc. (LeChevallier, 1991; Taylor et al., 2001) (Development of a model of BDOC and bacterial biomass fluctuation in distribution system[J], 1995,8(4)). Researches established the relationship between microbial regrowth and these parameters and discovered which circumstances biological stability parameter would restrict microbial regrowth. As for AOC, 10 μg/L is a universally acknowledged restriction concentration (Van der Kooij; Visser; Hijnen, 1982). Otherwise, in an actual water network, 50-100 μg/L may be acceptable to satisfy biological stability requirements (Laurent; Servais; Randon, 1993). However, whether or not the biological stability parameter could predict actual microbial regrowth in a complex distribution system is still unknown. At the same time, whether biological stability parameters should be adopted before or within the distribution is another question that remains answered (Lautenschlager et al., 2010, doi:; Hammes et al., 2008).

In this research, a waterworks and its distribution system were selected as the experimental object. In order to improve the water outlet quality, an advanced treatment process was added to the original conventional treatment process, which has comprised pre-ozonation and O_3-BAC process since Jan, 2010. Afterward, in Jan, 2011, its raw water was switched from the Huangpu River to the Qingcaosha Reservoir, taking water from Yangtze River. From Jul, 2009 to Nov, 2011, both the outlet and six successive pipeline sample points, in addition three secondary water supply sample points within the supply region, were selected as monitor sites.

Material and methods

The AOC sample bottles were treated with 2.5M nitric acid for 8 h, and washed with ultrapure water, then dried at 550°C for 2 h. The other sample bottles were washed with distilled water three times and dried at 105°C for 15 min. During this survey, temperature, residual chlorine and turbidity were measured online, and the other parameters were tested in the lab within 24 hours after sampling. The remained samples were kept at 4°C for seven days. Experiments involving indicators are listed as follows:

Chlorine: HACH46700-001 portable residual chlorine meter with DPD detection method. COD_{Mn} is determined by titration method of potassium permanganate with detection limits of 0.01 mg/L. The concentration of phosphate and ammonia nitrogen was obtained by the colorimetric method recommended by "analysis method for monitoring water and waste," the detection limit for which is 1 μg/L and 10 μg/L.

Total bacteria measurement: Take 1mL bacterial suspension, then dilute by $10, 10^2$ and 10^3 times. Both prime suspensions and dilutions were spread on the PCA culture medium plate, and bacteria numbers were counted after incubation at 37°C for 48 h in PCA culture medium composition: tryptone 10 g, beef extract 3 g, sodium chloride 5 g, agar 15 g, distilled water 1000 mL (P.R.C. National Standard, 2006).

Heterotrophic plate counts (HPCs) were counted on R2A agar plates which had been incubated for 7 days at 28°C. Duplicate samples measured and the results were expressed as CFU/mL (Uhl; Schaule, 2004).

Results

Bacterial regrowth in the water supply system with different performance

As Figure 1 shows, the bacteria in the secondary water supply system (SWSS) presented an obvious regeneration compared with the outlet and pipeline water. Adding advanced treatment can help reduce HPC in the pipeline and the secondary water supply system. Advanced water treatment combined with improved source water quality can greatly reduce bacterial regrowth in the water supply system, especially in the secondary water supply system. All these results showed that the source water quality played the most important role to control tap water quality.

Figure 1. HPC Comparison in the water supply System ("A" represents the period applying conventional process & using Huangpu River as raw water, "B" represents the advanced treatment & Huangpu River period, "C" represents the advanced treatment technology & Yangtze River period).

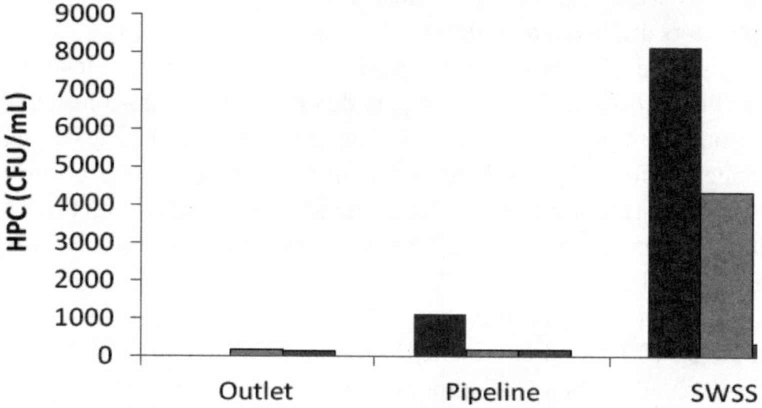

Some key water quality parameters in the water supply system

Table 1. Distribution system sampling survey results

	Process & Source water	R. Cl mg/L	Tur. NTU	HPC CFU/mL	TB CFU/mL	COD Mn mg/L	AOC µg/L	AN mg/L	TP µg/L
	Regular+Hungpu	1.50	0.11	27	1	3.45	178	0.72	54
WP	Advanced+Huangpu	1.77	0.20	181	1	2.18	91	0.59	5
	Advanced+Yangtze	1.54	0.15	98	1	1.38	73	0.42	2
	Regular+Hungpu	1.23	0.45	409	77	3.13	175	0.60	47
P.N.	Advanced+Huangpu	1.15	0.31	197	43	2.11	123	0.62	14
	Advanced+Yangtze	1.06	0.26	116	40	1.34	82	0.37	6
	Regular+Hungpu	0.39	0.68	6627	657	3.25	214	0.59	29
SWSS	Advanced+Huangpu	0.48	0.33	4354	462	2.27	116	0.50	35
	Advanced+Yangtze	0.76	0.33	365	69	1.45	73	0.41	16

Some key water quality parameters in the water supply system were measured. Table 1 shows the measurement for these parameters when the advanced treatment process was added, since Jan, 2010, and source water swtiched from Huangpu river to Changjiang river, since Jan, 2011.

Figure 2. AOC in the water supply system.

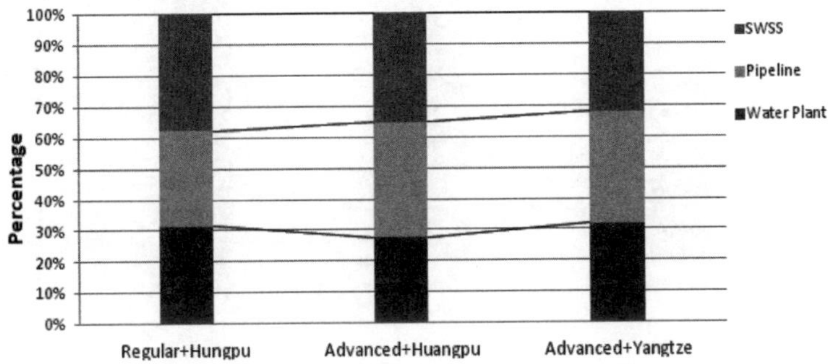

The AOC slightly fluctuated during distribution. During the regular treatment stage, the AOC in the secondary water supply system is close to or a little higher than the concentration in the pipeline and outlet; while in the next two phases, the AOC in the secondary water supply system is reduced when compared with the outlet water.

Table 1 shows that the AOC outlet continues decreasing after the advanced treatment process introduced and the raw water switched, while the pipeline network and secondary water supply system water quality have the same phenomenon.

Figure 3. Ammonia in the water supply system.

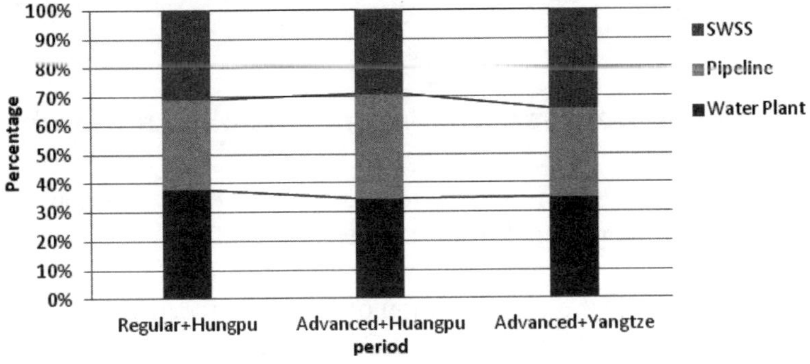

Ammonia in the distribution system had a different pattern of movement. Outlet water quality change does not cover ammonia, and the variation is not as significant as the other parameters. In the meanwhile, water distribution also does not affect ammonia a great deal.

Figure 4. TP in the water supply system.

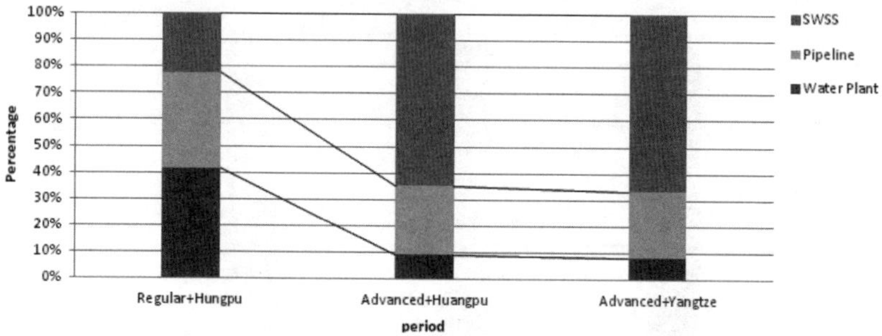

During the regular treatment process and using the Huangpu River as raw water stage, the total phosphorus decreased during distribution. However, after introducing the advanced treatment process, there was an increase of total phosphorus during the distribution. As a result, despite purification being significant in phosphate removal, the phosphate in the secondary water supply system might exceed the restrict limitation of microbial regrowth after water distribution and storage.

Nutrients in the water show different variations. AOC and ammonia did not change significantly, while the phosphate went up and down when the treatment process and the raw water were changed.

The concentration of AOC and TP was influenced by the process changes and water source switch. Ammonia was not affected by these changes.

Relationship between nutrient and microbial regrowth

As Figure 5 shows, if organic measured by AOC as a nutrient was the limiting factor, then 100 μg/L or lower would result in significant inhibition to microbial regrowth. AOC and HPC have a linear relationship while the concentration of AOC is under 150 μg/L, and HPC goes up with the increase of AOC.

Figure 5. The AOC to HPC curves.

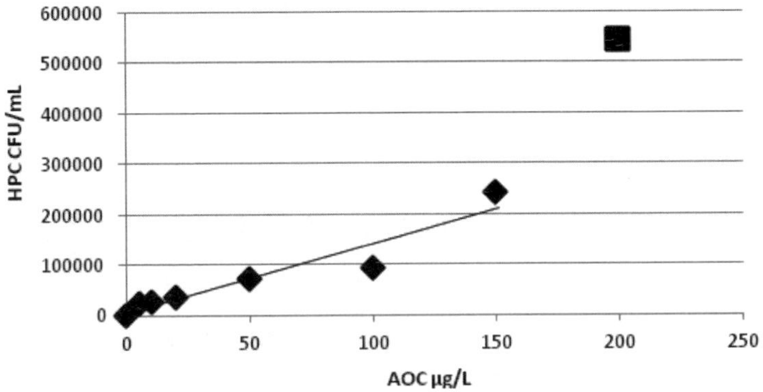

The correlation between the number of bacteria from the secondary water supply system and the water quality at the of water treatment plant outlet was analyzed using SPSS. The results are presented in Table 2.

Table 2. SWSS microbial and water outlet quality correlation

	T. Cl	Tur.	HPC	TB	COD_{Mn}	AOC	AN	TP
HPC	-0.067	0.284	0.322	0.373	0.525	0.667	0.201	0.518
TB	0.029	0.151	0.736	0.820	0.288	0.549	-0.094	0.003

Among these relationships, AOC has the best correlation with microbial number. The primary reason is that nitrogen and phosphorus concentration exceeded the restriction demands of microbial regrowth when compared with carbon source.

In a real water distribution system, the long residence time produces the good relationship between AOC and bacteria number, which is represented by HPC and total bacteria (Kerneys et al., 1995; Power; Nagy, 1999). In this research, the nutrient test revealed a certain linear relationship between AOC and the regeneration of the microbial number. As a result, if the HPC/total bacteria in tap water at 500/100 CFU/mL is taken as the thrust, and getting a linear fitter to outlet, then the AOC, AOC should be below 91 µg/L and 73 µg/L at the water works. This means that, in order to keep total bacteria below 100CFU/mL in the water distribution system, then the AOC at the outlet should be treated under 73 μg/L.

Discussion

The limiting factor to control bacterial regrowth in distribution system

Organic carbon, nitrogen and phosphate are all essential nutrients for bacteria growth. Insufficient amounts of these nutrients would influence bacteria regrowth. While in water distribution systems, not all nutrients meet the requirements to regulate bacteria regrowth.

First of all, nitrogen in the form of ammonia is not regulated by the purification process. As chloramine is a wildly used disinfectant in pipeline network, ammonia is around 0.2-0.4 mg/L in the water, and it is far beyond the limitation requirement for most of the bacteria (Volker; Schreiber; Kistemann, 2010).

Ammonia was always largely wiped out by the bacterial activities during BAC filtration. But this removal efficiency is affected by seasonal temperature variation. As a matter of fact, ammonia is hard to maintain at a fairly acceptable lower concentration for restricting microbial regrowth (Lipponen; Suutari; Martikainen, 2002).

Phosphate has the same disadvantage as nitrogen. Although phosphate can be removed at 2 μg/L in the water treatment process, this obviously increasing trend is observed in the distribution system. This is perhaps relevant to the release from the water pipe. Meanwhile, phosphate showed less correlation compared to the AOC (Srinivasan et al., 2008).

However, AOC at the water works outlet showed some correlation with the microbial concentration in the tap water. In the water supply system, AOC is the only practicable nutrient parameter to restrict microbial regrowth (Assimilable organic carbon(AOC) in drinking water[C], 1990).

Bacteria regrowth is a common phenomenon in water distribution systems. However, whether or not bacteria regrowth is out of control should be judged by some standards. The standard for total bacteria is 100 CFU/mL. The standard for HPC is 500 CFU/mL (USEPA, 2002). It was very difficult for the water works to completely prevent microbial regrowth, but it was possible to restrain the regrowth at a certain level.

Tap water is the most concerning point (Lautenschlager et al., 2010, doi:), and microbial regrowth mainly took place at secondary water supply system. In this research, over two thirds of detected bacteria were reproduced at the secondary water supply system. Biological stability stands for the maximum possibility for microbial regrowth; as a result, the biological stability parameter will have a closer relation with bacteria number in tap water (Dukan, 1996; Huck; Gagnon, 2004).

Limiting factor to control bacteria regrowth in pipeline

As shown in Table 1, when AOC at the water works outlet is above 93 μg/L, microbial number would exceed the standard. For instance, in 2009, when the AOC outlet is 178 μg/L, the HPC for tap water is 6627CFU/mL on average; in 2011, when the AOC outlet is 73 μg/L, the HPC tap water is 365CFU/mL on average. All these results proved that, in this water supply system, AOC can be the limiting factor and can be calculated to provide the correct value for water treatment plants.

Conclusion

Water quality in the distribution system can be obviously improved. Total bacterial and HPC in the water distribution system were reduced after introducing O_3-BAC in to the advanced treatment process and switching water source.

Organic carbon can be the only limiting factor for microbial regrowth control in water distribution systems. In order to keep the total bacterial below 100CFU/mL in the water distribution system, AOC at the water works outlet should be treated under 73 μg/L.

References

DAVID, A. A.; MOSSEL, B. StruijkCorry. Assessment of the microbial integrity, sensu G.S. Wilson, of piped and bottled drinking water in the condition as ingested. International Journal of Food Microbiology, 92, p.375-90, 2004.

DUKAN, S. Dynamic modelling of bacterial growth in drinking water networks. Water Research, 30, p.1991-2002, 1996.

GHIZELLAOUI, S. Evaluation and evolution of the quality of the water resources in the distribution network. Desalination, 222, p.502-12, 2008.

HAMMES, F. et al. Flow-cytometric total bacterial cell counts as a descriptive microbiological parameter for drinking water treatment processes. Water Research, 42, p.269-77, 2008.

HAMMES, F. et al. Formation of assimilable organic carbon (AOC) and specific natural organic matter (NOM) fractions during ozonation of phytoplankton. Water Research, 41, p.1447-54, 2007.

HUCK, P. M.; GAGNON, G. A. Understanding the distribution system as a bioreactor: a framework for managing heterotrophic plate count levels. International Journal of Food Microbiology, 92(3), p.347-53, 2004.

KERNEYS, A. et al. The effects of water residence time on the biological quality in a distribution network. Water Research, 29(7), p.1719-27, 1995.

KISTEMANN, T.; DANGENDORF, F.; EXNER, M. A Geographical Information System (GIS) as a tool for microbial risk assessment in catchment areas of drinking water reservoirs. Int. J. Hyg. Environ. Health, 203, p.225-33, 2001.

LAURENT, P.; SERVAIS, P.; RANDON, G. Bacterial development in distribution networks-study and modelling. J. of Water Supply, 11, p.387-98, 1993.

LECHEVALLIER, M. W. Bacterial nutrients in drinking water. Appl. Environ. Microbiol, 53(3), p.857-62, 1991.

LECHEVALLIER, M. W. Coliform regrowth in drinking water: a review. J AWWA, 82(11), p.74-86, 1990.

LECHEVALLIER, M. W. et al. Full-scale studies of factors related to coliform regrowth in drinking water. Applied and Environmental Microbiology, 52(7), p.2201-11, 1996.

LIPPONEN, M. T. T.; SUUTARI, M. H.; MARTIKAINEN, P. J. Occurrence of nitrifying bacteria and nitrification in Finnish drinking water distribution systems[J]. Water Research, 36, p.4319-29, 2002.

POWER, K. N.; NAGY, L. A. Relationship between bacterial regrowth and some physical and chemical parameters within Sydney's drinking water distribution system. Water Research, 33, p.741-50, 1999.

SERVAIS, P. Development of a model of BDOC and bacterial biomass fluctuation in distribution system. Journal of Water Science, 8(4), p.427-62, 1995.

SRINIVASAN, S. et al. Factors affecting bulk to total bacteria ratio in drinking water distribution systems. Water Research, 42, p.3393-404, 2008.

TAYLOR, I. Bacterial growth in distribution system: Effect of assimilable ogranic carbon and biodegradable dissolved organic carbon. Environment Science Technology, 35(27), p.3442-7, 2001.

UHL, W.; SCHAULE, G. Establishment of HPC(R2A) for regrowth control in non-chlorinated distribution systems. International Journal of Food Microbiology, 92, p.317-25, 2004.

USEPA. National Primary Drinking Water Regulations, 2002.

VAN DER KOOIJ, D. Assimilable organic carbon(AOC) in drinking water. In: Drinking water microbiology. New York: Springer-Verlag, 1990. p.57-87.

VAN DER KOOIJ, D.; VISSER, A.; HIJNEN, W. A. M. Determining the concentration of easily assimilable organic carbon in drinking water. J Am Water Works Assoc., 74(10), p.540-5, 1982.

VOLKER, S.; SCHREIBER, C.; KISTEMANN, T. Drinking water quality in household supply infrastructure-A survey of the current situation in Germany. International Journal of Hygiene and Environmental Health, 213, p.204-9, 2010.

16

Measurement of Mercury in Flue Gas Based on an Aluminum Matrix Sorbent and Mercury Removal by Existing Pollutant Control Devices of Six Coal-fired Power Plants in China

Wang Wenhua

Introduction

Mercury is considered to be a global pollutant in the atmosphere due to its toxicity, volatility, persistence and long range transport capability. Global Hg emissions for 2000 were reviewed in detail by Pacyna et al. (2006) indicating that two thirds (1.46 kt) of total atmospheric emissions (2.19 kt) originated from coal combustion. Mercury emissions from coal-fired flue gas has aroused increasing attention and many efforts have been made to develop strategies and technologies to control such emission.

At present, there are quite a few methods applied in sampling and quantifying Hg in flue gases. Several EPA standard procedures (methods 29, 101A and OH method) have been widely recommended to determine the particulate and gaseous mercury emissions from stationary sources (Usepa, 1982; Usepa, 1992; ASTM, 2002). These EPA methods are based on isokinetic sampling of the flue gases at a suitable sampling point using a filter collection system for the particulates. The sampled gases then pass through a set of impingers containing liquid absorbents such as KCl, HNO_3/H_2O_2 and $KMnO_4/H_2SO_4$ to trap all the present mercury in the sample gases. Some other aqueous oxidative solutions were tested by Shendikar, Dample and Gutknecht (1984), such as 10% H_2O_2, acidified 10% H_2O_2, 0.2 M $(NH_4)_2S_2O_8$, 0.2 M $(NH_4)_2S_2O_8$ + 0.025 M $AgNO_3$, and 1.5% $KMnO_4$ in 10% H_2SO_4. With the last two mixtures, good collection efficiencies for Hg^0 were exclusively obtained when using more than one impinger. $K_2Cr_2O_7$ (4%) in 4 M HNO_3 solution was used to sample Hg in flue gas of a coal fired power plant by Meij (1991).

Meanwhile, many solid mediums were also used to collect Hg from flue gas. In Frontier Geosciences Mercury Speciation Method (Prestbo; Bloom, 1995), flue gas passes through a set of two KCl/soda lime traps and two iodated carbon traps. Hg^{2+} and Hg^0 are collected by the two different sorbents, respectively. In MIT's Solid Sorbent Method (Olmez; Aras, 1993), flue gas passes through two activated charcoal sorbent traps connected in proper order, which capture Hg^{2+} as well as Hg^0 at the same time. Using the HEST Method (Cooper, 1994), carbon impregnated filters are used to adsorb gas phase Hg while activated carbon, treated with iodine or halogen, has been applied to capture total Hg in flue gas using EPA method 30B (Usepa, 2007) and method Appendix K (40 CFR…, 2007).

These methods could be classified into two main groups: methods to employ impingers with oxidant solutions and methods based on solid sorbents. Although the solid sorbent method is easier to handle than the liquid sorbent method, researchers found that the collective efficiency of Hg^0 was decreased along with the increasing temperature of flue gas and the increasing level of SO_2 in most of the solid sorbents (Meij; Spoelstra; Waard, 1989; Zhuang et al., 2007; Zhao et al., 2000). Thus, a number of the disadvantages found in the methods based on solid sorbents have to be improved.

In our previous study (Mei et al., 2008a; Mei et al., 2008b), a series of aluminum-matrix-based sorbents were prepared to absorb Hg by converting Hg^0 to Hg^{2+} with catalytic oxidation. Characteristics of the sorbents, such as the size of the particles, crystal structures, binding energies, Brunauer–Emmett–Teller (BET) specific surface area (S_{BET}) and pore size distribution, were respectively analyzed. The catalytic and adsorption tests were carried out by lab testing with simulated flue gas (consisted of N_2, O_2, SO_2, NO and HCl). These sorbents showed a good SO_2 anti-poisoning ability and stable oxidation and adsorption abilities in a wide temperature range. The sorbents could be easily regenerated by thermal decomposition above 500 °C. This feature made the application of these sorbents more economical than carbon-based sorbents, especially in China.

Recently, the rapid economic and industrial developments in China have led to the dramatic increase in power demand and coal consumption. It was estimated that the amount of coal consumed in China (1.27×10^9 ton) was about 21% of the world's total coal consumption in 1990, which has increased to 25% in 2001, and will be 35% in 2015 (Koukouzas et al., 2006). In 2003, the total power generation capacity has reached 391 GW, among which about 70% came from burning coal (Mukherjee et al., 2008). It was speculated that the thermal power sector would utilize 73% of the coal consumed in China in 2030 (Koukouzas et al., 2006). Therefore, China is considered to be one of the major sources of global anthropogenic mercury emissions.

Mercury emission from Chinese stationary coal-fired power plants was estimated to be 213.5 tons in 1995 (Wang et al., 1996), 202.4 tons in 1999 (Streets et al., 2005), and 220 tons in 2000 (Jiang et al., 2005). According to the report of the 25th session of the Governing Council/Global Ministerial Environment Forum (Unep, 2009), global atmospheric emissions of mercury from human activities in 2005 were estimated to be approximately 1930 (ranging from 1230 to 2890) tons. China is considered to be the largest contributor with about 825 tons, 45% of which are emitted due to the burning of fossil fuels (primarily coal) (Unep, 2009). However, there are many uncertainties (up to ± 40%) in this estimate due to the lack of data from actual source measurements in China, such as the mercury as well as other components' concentration in the coal or insufficient technologies both for productive means and mercury pollution control.

In China, most of the coal-fired power plants have been equipped with both the cold-side electrostatic precipitator (ESP) and wet flue gas desulfurization (WFGD) procedure to remove the particles and SO2, respectively. Additionally, the selective catalytic reduction (SCR) system has been installed in some plants to control nitrogen oxides (NOx). Previous studies indicate that SCR, ESP and FGD (Flue Gas Desulfurization) can affect the mercury speciation and remove part of the mercury from the flue gas. Reviews of EPA's Information

Collection Request (ICR) data on mercury captured in boilers and existing air pollutant control devices (APCDs) indicate that mercury removals through cold-side ESP approach could be up to 27% while only 4% via hot-side ESP. Both the wet and dry FGD systems can remove 80% to 90% of the gaseous mercury (Hg^{2+}) with no effect on elemental mercury (Hg^0) removal (Pavlish et al., 2003). Richardson et al. (2002) indicated Hg^0 could be oxidized to Hg^{2+} on the surface of an SCR catalyst with a high catalytic oxidation efficiency between 80% and 90%. The generated Hg^{2+} could be subsequently absorbed by the wet slurry in FGD system. However, there are few research projects on mercury control by existing APCDs and mercury emission factors for Chinese power plants, which leads to an erroneously estimated value of mercury emission in China.

In this study, the sorbent AL-CCl ($CuCoO_4$ + NH_4Cl loaded on γ-Al_2O_3) was selected as the collection medium to capture Hg in flue gas. A sampling and detection method was established with a small system for coal combustion and flue gas sampling in the lab. Six coal-fired power plants in China were selected to practice onsite mercury measurements, investigate the mercury removal by existing APCDs and estimate the fate of mercury from coal-fired power plants. All the gaseous and solid samples, including pulverized coal, bottom ash, flying ash captured in ESP and gypsum in the FGD, were collected at each coal-fired power plant. The Hg in flue gas was trapped with an aluminum matrix sorbent and its emissions and removals were estimated. The overall mercury mass balances and mercury emission factors were discussed based on the data in each power plant.

Experiment

Materials and Determine Method

$CuCoO_4$ + NH_4Cl loaded on γ-Al_2O_3 (AL-CCl) was prepared by impregnating AL-C ($CuCoO_4$ loaded on γ-Al_2O_3) with 30 wt% NH_4Cl solutions before being decomposed and dried at 60 °C for 8 hours, and then calcined at 400 °C for 3 h. AL-C was prepared by the thermal decomposition of Cu–nitrate/Co–nitrate mixtures loaded on

γ-Al_2O_3. Both nitrates were mixed in Co/Cu with an atomic ratio of 1. The mixtures were dissolved in water to form the mixtures. γ-Al_2O_3 was impregnated by the mixtures solution in proportion to 20 wt% nitrate loading values and heated to 230 °C, the temperature at which the nitrogen oxides were evolved. The heterogeneous mass obtained was ground in a mortar and heated for 5 h at 400 °C.

A coal sample from south-west China was used in a research lab after a pretreatment of grinding and sieving. They were digested in a V_2O_5-HNO_3-H_2SO_4 solution and then the solution was analyzed by a cold vapor atomic adsorption spectrometry mercury analyzer (CVAAS, SG-921, Jiangfen Ltd.). A standard reference material (BCR-40) with a certified value for mercury (0.35 ± 0.06 mg/kg) was introduced for quality control. All the chemicals were of analytical grades. The major components of BCR-40 and coal are listed in Table 1.

Table 1. Major components of BCR-40 and coal sample (dry air basis)

Coal type	Moisture (%)	Ash (%)	Volatile (%)	N (%)	C (%)	S (%)	Cl (%)	[Hg]mean (mg/kg)
BCR-40	1.20	16.1	13.0	1.20	70.6	1.28	0.230	0.354 ± 0.003
Coal	2.97	10.4	13.2	1.03	83.1	1.41	0.110	4.61 ± 0.04

The mercury concentration of BCR-40 was detected at an average value of 0.354 ± 0.003 mg/kg ($n = 5$) with a relative error < 2% compared to certified value. Therefore, the determination course was reliable.

Sample Collection

Mercury in flue gas was collected using a sorbent trap with three tubes connected by Teflon sleeves. All the tubes were made by quartz (50 mm in length with an outer diameter of 10 mm and an inner diameter of 7.5 mm) containing approximately 1 g of sorbent which was previously sandwiched and fixed by the quartz cotton. Among the three tubes, tube 1 was designated for primary capture of gaseous Hg. Tube 2 was designated as a backup to check vapor phase Hg^0 breakthrough. The third tube, with a known mass of Hg^0, was spiked to perform the analytical bias test and the spiked Hg^0 recovery study. Each sorbent trap was marked with an identified number for tracking.

During the lab research, the mercury collection experiment was carried using coal combustion and sampling equipment, as shown in Figure 1. A porcelain boat with the coal was put into a quartz tube (650 mm in length with an outer diameter of 36 mm and an inner diameter of 32 mm), which crossed through a clam-shell furnace. Figure 2 presents that Hg emission was started from 150 °C in a furnace and the maximum emission was reached during 240-250 °C. Most of the volatile mercury in the coal was released into flue gas at 600 °C (Figure 3). Therefore, the temperature in the furnace was maintained at 600 °C. The gas pump was turned on before trapping Hg, and the gas flow was adjusted to 0.3-0.5 L/min. With the setup, coal-fired flue gases were extracted through the sorbent trap. The sampling time was 5 minutes. The CVAAS at the end of the equipment, as shown in Figure 1, was connected in order to monitor mercury loss in the gas path. After trapping, the mercury content in the coal ash was tested with the same procedure as with the coal, and the average mercury concentration (1.50 ± 0.11 mg/kg, $n = 15$) was obtained

Field mercury emission measurements based on this method were practiced in six coal-fired power plants in China. Figure 4 presents the sampling sites in the flue of all the six power plants tested in this study.

As shown in Figure 5, the test gases were sampled from the flue gas stream at an appropriate flow rate (0.3-0.5 L/min) through a heated probe that prevented water condensation occuring before the gas entered the sorbent trap(s). A water knockout after the sorbent trap was placed to prevent the water in the sample line from entering the flue gas sampler (TH-600, Tianhong, China). The samples were simultaneously taken from both the inlet and outlet of the pollutant control devices, e.g., ESP and WFGD, as shown in Figure 2. According to the results of the pretest, 2 hours were enough for the gas phase mercury detection. After the sampling, each tube of the sorbent trap was immediately placed into Hg-free containers for subsequent laboratory detection. The characteristics (such as gas temperature, static pressure and gas velocity) and main components (such as SO_2, NO_x and O_2) of the flue gas were determined by the flue gas analyzer (350-XL, Testo, German) during the sampling period. Solid samples including the pulverized coal, bottom ash, fly ash captured in ESP and gypsum in the WFGD, were taken after half-hour of Hg trapping. The sampling was conducted two or three times for each power plant to obtain the parallel results and reduce uncertainties. The data obtained from the tests using the same conditions were averaged to get the final results.

Figure 1. Schematic diagram of the coal combustion and sampling system.

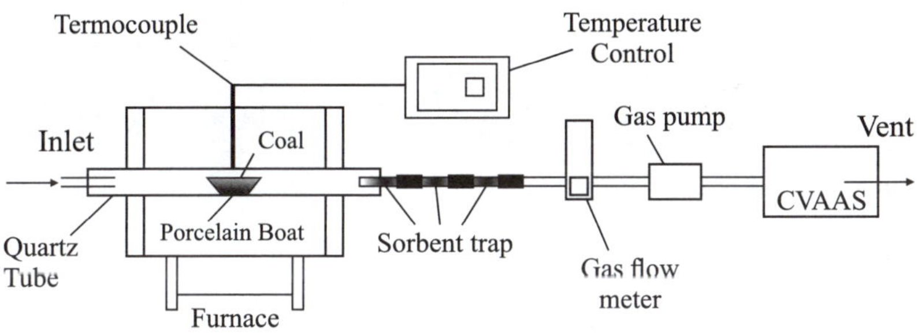

Figure 2. Hg emission from coal with the increased temperature in furnace.

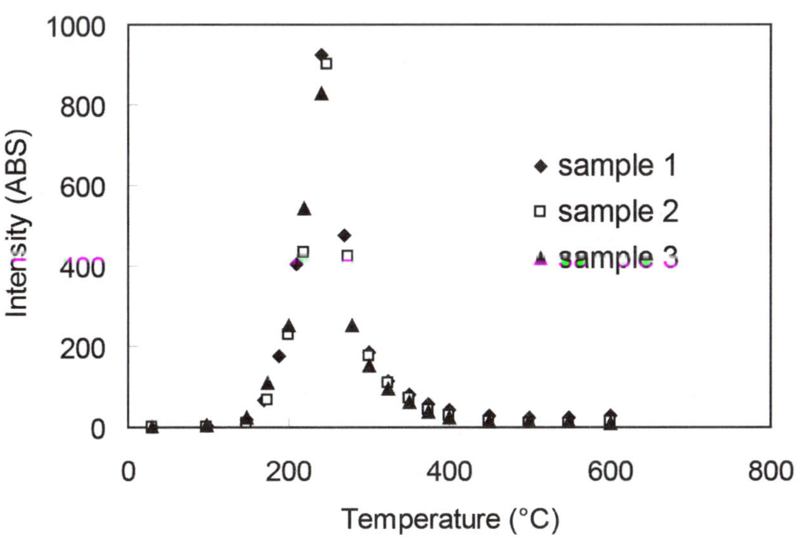

Figure 3. Hg emission from coal with and without sorbent trap at 600 °C.

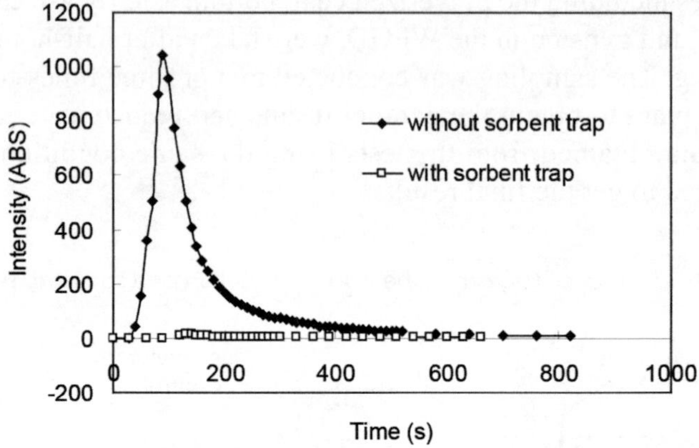

Figure 4. Simultaneous power plant sampling locations.

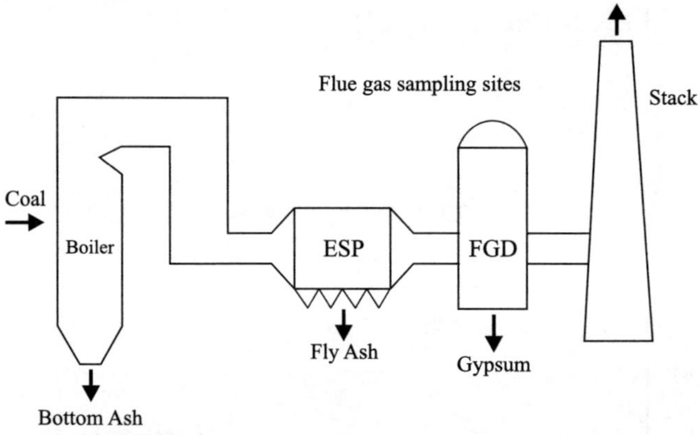

Figure 5. Sampling train in field.

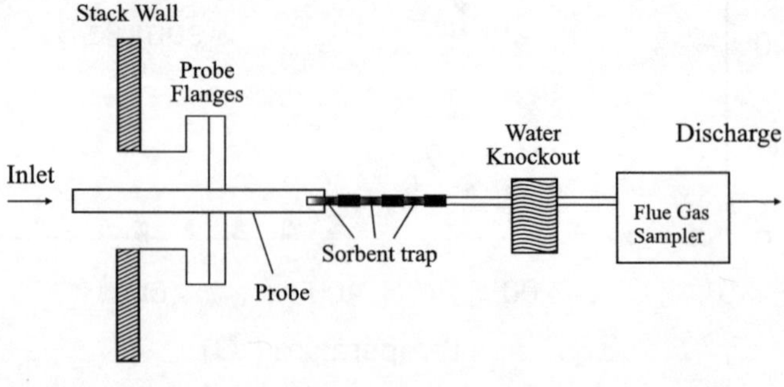

Detection of Hg in Flue Gases

The individual tube of the sorbent trap and their respective components were detected respectively. Sorbent media, quartz cotton and fly ash deposited on the quartz cotton in each tube were moved into a glass funnel covered with a piece of filter paper. The samples were eluted by 50 mL 4 mol/L HCl - saturated KCl solution, in which the captured mercury was transferred into liquid solution. The mercury concentrations were measured by CVAAS (SG-921, Jiangfen, China). The amount of Hg contents in tube 1 and tube 2 was the total mercury concentration in the flue gas. From the results detected in tube 3, spiked with a known mass of Hg^0, the recovery and performance of the combined sampling and analytical approach could be verified. The solid samples taken from the power plants were measured by a Zeeman Mercury Spectrometer (RA-915$^+$, Lumex, Russia). All the instruments were calibrated before the sample analyses.

Results and Discussion

Recovery of Hg Analysis in Lab

Recovery of spiked Hg0 in tube 3 of the sorbent traps

The ability to recover and accurately quantify Hg^0 on the sorbent media was evaluated by performing an analytical bias test on tube 3 spiked with Hg^0. Mercury concentrations in tube 3 were analyzed using the same procedure as mentioned above. In Table 2, the average Hg^0 recovery of 97.3% in tube 3 was measured without coal combustion. The specific verification of the performance of the combined sampling and detection approach was also provided by performing the spiked Hg^0 recovery test. As shown in Table 3, the average Hg^0 recovery of tube 3 varied from 89.7% to 105% (average of 97.0%) with various coal consumption levels. The above results demonstrate that the sorbent media in tube 1 and tube 2 possessed efficient capacity of mercury adsorption and there was no gas phase Hg breakthrough. Spiked Hg^0 in tube 3 was not affected by the passing-by flue gas. The approach of sampling and detection in this study was validated in summary.

Table 2. Recovery of spiked Hg^0 in tube 3 in lab test (n = 3).

Coal consumption (g)	Hg emission from coal (μg)	Hg0 spiked in tube 3 (μg)	Hg0 recovered from tube 3 (μg)	Average recovery (%)
0	0	0.25	0.243 ± 0.002	97.3
1.0	3.11	0.25	0.247 ± 0.012	98.7
2.0	6.22	0.25	0.241 ± 0.020	96.2
3.0	9.33	0.25	0.224 ± 0.003	89.7
4.0	12.44	0.25	0.263 ± 0.041	105
5.0	15.55	0.25	0.238 ± 0.036	95.4

Hg emission from coal (μg) = (4.61 – 1.50) (μg/g) × coal consumption (g), Average recovery (%) = [Hg^0 recovered from tube 3 (μg) ÷ 0.25 (μg)] × 100%.

Recovery of Hg in flue gas

Hg recovery efficiency in tube 1 and tube 2 of the traps after sampling were calculated to evaluate the feasibility and accuracy of this method. In Table 3, the Hg content in original sorbent media before sampling was under the detection limit of our method, which indicated that Hg was totally removed from the sorbent during the baking procedure. Hg contents in tube 1 increased with coal consumption (Figure 6), the linear correlation coefficient of which is quite good (i.e., R^2 = 0.998). However, Hg contents in tube 2 were negligible compared to that of tube 1. The average recovery of tube 1 and tube 2 varied from 83.0% to 94.3% (average of 87.7%). Due to the complex composition of the flue gas, the adsorption of Hg by the sorbent was affected by many factors, such as fluctuations in coal, ash, flue gas flow rates, sampling uncertainty and measurement uncertainty. Yokoyama et al. (2000) considered that results between 70 and 130% were acceptable due to the difficulties in estimating each uncertainty.

Table 3. Recovery between Hg trapped in flue gas and Hg emission from coal (n = 3).

Coal consumption (g)	Hg emission from coal (μg)	Hg trapped in tube 1 (μg)	Hg trapped in tube 2 (μg)	Average recovery (%)
0	0	Not detected	Not detected	Not detected
1.0	3.11	2.92 ± 0.03	0.017 ± 0.003	94.3

Coal consumption (g)	Hg emission from coal (μg)	Hg trapped in tube 1 (μg)	Hg trapped in tube 2 (μg)	Average recovery (%)
2.0	6.22	5.56 ± 0.10	0.052 ± 0.000	90.3
3.0	9.33	7.72 ± 0.16	0.023 ± 0.002	83.0
4.0	12.44	10.8 ± 0.2	0.026 ± 0.003	86.9
5.0	15.55	13.0 ± 0.6	0.036 ± 0.004	84.0

Average recovery (%) = [Hg recovered form tube 1 and tube 2 (*μ*g) ÷ Hg emission from coal (*μ*g)] × 100%.

Figure 6. Distribution of Hg in tube 1 and tube 2.

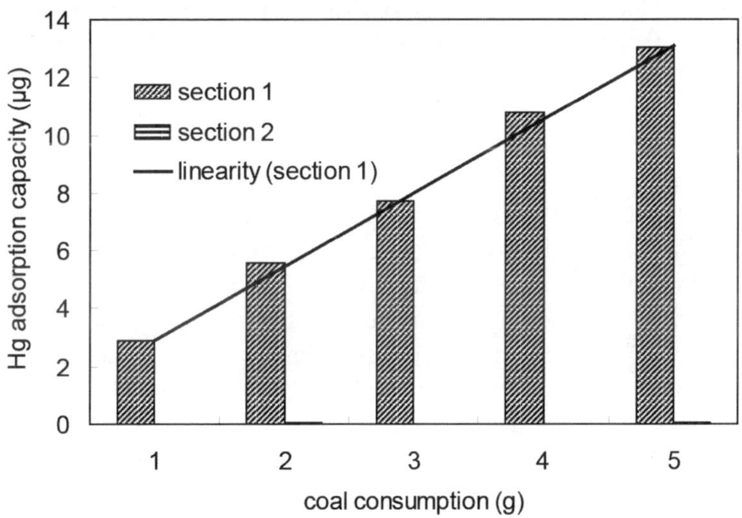

The Practice of the Analysis Method in Power Plants

In order to evaluate the efficiency of this method, the whole processes of field mercury emission measurements had to be tested in six coal-fired power plants in China. The data of from the six power plants were summarized in Table 4. Four of the six power plants had pulverized coal power generation units with different capacities, while the other two had advanced supercritical boilers. All the six power plants were equipped with tangential burners, cold-side ESP and wet FGD systems, where the watered limestone slurry was used to remove sulfur oxides and gypsum was produced as the byproduct. The fourth plant had an additional SCR system.

Table 4. The data of the six power plants.

Power plant no.	Boiler type	Coal type	Capacity (Mw)	Pollution Control Device
1	Pulverized coal	bituminous coal	98	ESP + WFGD
2	Pulverized coal	bituminous coal	300	ESP + WFGD
3	Supercritical	bituminous coal	758	ESP + WFGD
4	Supercritical	bituminous coal	1000	SCR + ESP + WFGD
5	Pulverized coal	bituminous coal	100	ESP + WFGD
6	Pulverized coal	Anthracite coal	162	ESP + WFGD

Coal analyses

The annual coal consumption by the electric utilities in China was approximately 1.6×10^9 ton. Ren et al. (2006) compiled Hg data of 1413 coal samples and calculated the average Hg content of 0.20 mg/kg. Zhang et al. (1999) reported an average Hg content in Chinese coals of 0.16 mg/kg based on 990 coal samples, which is higher than the world average (0.1 mg/kg) and slightly lower than the U.S. average (0.17 mg/kg) (Zheng; Liu; Chou, 2007). Wang et al. (1996) compiled literature data on Hg content in Chinese coals and calculated an average Hg content of 0.22 mg/kg.

The coals burned in first five plants came from the north China while the coal from the last plant came from a Chinese mine whose Hg content was higher than the coal from the other area. Proximate and ultimate analyses of the burned coals are shown in Table 5.

Table 5. Proximate and ultimate analyses of the burned coals (dry air basis).

Component	Coal-1	Coal-2	Coal-3	Coal-4	Coal-5	Coal-6
Volatile (%)	37.3	32.2	34.9	31.2	33.0	9.86
Ash (%)	31.6	16.2	11.6	12.5	23.4	36.9

Component	Coal-1	Coal-2	Coal-3	Coal-4	Coal-5	Coal-6
Moisture (%)	8,50	12.3	3.43	14.3	4.40	1.88
(%)	0.570	0.43	0.540	0.38	0.920	2.41
S (%)	0.790	0.98	0.860	1.03	0.990	1.15
N (%)	55.3	65.5	49.4	69.9	54.4	61.4
C (%)	**628**	**538**	**472**	**573**	**663**	**940**
Cl (mg/kg)	0.161 ±	0.147 ±	0.0640 ±	0.088 ±	0.225 ±	0.320 ±
Hg (mg/kg)	0.029	0.038	0.0035	0.010	0.054	0.091

Mercury mass balances

Mercury mass balances were estimated, the results from which are shown in Table 6. The operation parameters, such as coal/lime feeding rate, bottom ash production rate, fly ash and gypsum generation rate, were provided by the power plants.

The mercury mass balances of the six power plants were 91.6%, 75.4%, 77.1%, 74.3%, 118% and 85.8%, respectively. There were some uncertainties that may have affected the accuracy of the mass balance estimation, which were mainly caused by the inhomogeneity of the components in the coal/ash/gypsum, fluctuation of boiler operating conditions and the uncertainty of sampling and chemical analysis procedures. Previous research suggested that the results were acceptable if the mass balance closures were between 70% and 130% (Yokoyama et al., 2000). Results from the six power plants were within this range. Some other previous investigations have relatively similar ranges of recovery. Zhang et al. (2008) calculated the Hg mass balances of six power plants, which varied from 100.3% to 139.5%. Yokoyama et al. (2000) estimated the mass balances of Hg on three type coals between 89% and 138%. The recovery rates of 82% and 99% were referred to by Lee et al. (2006).

Table 6. Mass balances of Hg between input and output streams in four power plants (*n* = 3).

Stream			Plant-1	Plant-2	Plant-3	Plant-4	Plant-5	Plant-6
Input	Coal	g/hr	4.48	18.7	16.9	24.5	8.24	18.1
Output	Bottom ash	g/hr	5.00x 10^{-3}	76.0x 10^{-3}	2.00× 10^{-3}	0	3.00× 10^{-3}	452× 10^{-3}

	Stream		Plant-1	Plant-2	Plant-3	Plant-4	Plant-5	Plant-6
	Fly ash	g/hr	0.193	0	0.366	0. 425	0.721	4.23
	Gypsum	g/hr	2.60	0.558	0.373	2.01	7.18	3.84
	Stack flue gas	g/hr	1.31	13.5	12.3	15.8	1.79	7.02
Mass balance	(Input/ output) ×100	%	91.6	75.4	77.1	74.3	118	85.8

Distribution of mercury

During the process of coal stream in the power plant, the mercury in the coals is distributed into bottom ash, ESP ash (fly ash), FGD reactant (gypsum) and flue gas. The relative distribution of Hg and the Hg concentrations in solid/gas samples in four power plants are summarized in Table 7. These values were the average of the data which varied depending on factors such as coal type and the operational conditions of the plants.

Table 7. Relative distribution of Hg and the mean concentrations of Hg in solid/gas samples in four power plants (*n* = 3).

Plant no.		1	2	3	4	5	6
Distribution (%)	Botton ash	0.110	0.410	1.00x 10^{-2}	0	4.00x 10^{-2}	2.50
	ESP	4.30	0	2.17	1.74	8.75	23.4
	WFGD	58.0	2.98	2.21	8.20	87.1	21.2
	Stack	29.2	72.2	72.7	64.5	21.8	38.8
Concentration (μg/kg or μg/m³)	Bottom ash	6.96± 1.16	4.77± 0.88	0.420± 0.996	8.50± 2.79	3.25± 3.86	49.3± 2.62
	Fly ash	30.6± 5.50	0	8.50± 1.31	502± 24.7	92.2± 28.6	68.5± 29.0
	Gypsum	156± 6.46	250± 19.2	6.22± 1.86	4.59± 0.84	1560± 74.3	231± 64.9
	Stack flue gas	2.27± 0.58	8.47± 0.97	4.29± 0.45		1.56± 0.68	5.95± 1.80

Ranging from 0% to 2.50%, little mercury was found in the bottom ashes which is consistent with previous investigations (Yokoyama et al., 2000; Lee et al., 2006; Meij; Vredendregt; Winkel, 2002). The mercury concentration in the bottom ash of plant-6 was far higher than those of the other five plants due to the difference of coal types. The coal burned in plant-6 was anthracite coal with low volatility (< 10%) and those in the first five plants were bituminous coals with higher volatile (20%--40%). Anthracite coal is easy to be slagged and agglomerated during combustion and is covered by melted inclusions which would deteriorate its burn out capacity (Du; Zhang, 1994). Therefore, incompletely burning the coal could not have lead to the thorough emission of the mercury.

The relative distribution of Hg in ESP ranged from 0% to 23.4%. During coal combustion, the majority of the mercury was evaporated in the boiler. When the temperature decreases, gaseous mercury could be condensed or adsorbed on the surface of the fly ash particles which were collected in the ESP afterwards (Zhang et al., 2008). In plant-6, about one quarter of Hg (23.4%) was removed through the ESP which showed much more efficiency than the others. A high concentration of Hg in the coal and a large production rate of fly ash might be the major reasons for this diversity.

Some oxidized mercury in flue gas could be absorbed by lime-slurry in the FGD scrubber tower due to its water-soluble property (Brown et al., 1999). The relative distribution of Hg in FGD for the six power plants ranged from 2.21% to 87.1%, which varies with coal types, tower types and liquid-to-gas (L/G) ratios. Hg^0 could be converted into Hg^{2+} by reacting with chlorine ions in flue gas (Hall; Schager; Lindqvist, 1991). As shown in Table 5, the concentration of Cl^- in coal-1 was similar to that in coal-5. However, the fraction of Hg in FGD in plant 1 is much lower than that of plant 5, which indicates that the Cl^- concentration in the coal may not be the major factor to control the Hg removal efficiency of FGD. Although the highest concentrations of Hg and Cl^- are found in coal-6, the fraction of Hg in FGD in plant 6 was lower than the average level (29.9%) of all six plants. The highest concentrations of SO_2 (2.41%) in coal-6 may be the important reason. The oxidation of Hg^0 would be restrained by SO_2 in flue gas in the presence of HCl (Ghorishi et al., 2005).

In stacks, the distribution of Hg ranged from 21.8% to 72.7%, and the final emissions into the atmosphere from the stacks ranged from 1.56–8.47 μg/m^3. Compared with other studies, such as Guo; Zheng; Xu (2007) with a range of 13--21μg/m^3 (with 0.33 mg/kg Hg in coal), Otero-Rey et al. (2003) with a range of 19.1--21.8μg/m^3 (with 0.15--0.17 mg/kg Hg in coal) and Lee et al. (2006) with a range of 1.03--2.41μg/m^3 (with 0.043--0.078 mg/kg Hg in coal), Hg emissions from the six power plants in this study were relatively low.

Mercury removals by existing air pollutants control devices

The above results showed that current air pollutant control devices (cold-side ESP and wet FGD) in power plants can remove part of the Hg from flue gas. The average mercury removal efficiency of the ESP+FGD systems is 44.2% in this study, which corresponds with the US EPA results that ranged from 29% to 74% (US EPA, 1999; US EPA, 2002a; US EPA, 2002b). The ESP+FGD or fabric filter (FF) +FGD systems have been applied in most of the power plants in China recently. Considerable numbers of fluidized bed combustion (FBC) boilers have been applied for power generation, which remarkably promote the adsorption of Hg by fly ash. Mercury control technology in flue gas are being investigated and developed. Although the coal consumption in power industry increases annually, these APCDs may play an important role in the reduction of mercury emissions, which should have been considered when total Hg emission from power plants was recently estimated for China.

Mercury emission factors

The mercury emission factor is a very important parameter to estimate total mercury emissions. Pacyna et al. (2006) estimated global mercury emissions in 2000 from power plants with the factor 0.1--0.3 g/ton, and the emission factor of Chinese power plants in 1999 calculated by Streets et al. (2005) was 0.136 g/ton. In this study, the mercury emission factors were calculated as 0.045--0.124 g/ton and with an average of 0.071 g/ton for the six power plants, which was much lower than the results mentioned above. This difference can be assigned to the installation of APCDs in Chinese power plants being in place since 2000. In the past decade, the newly-built power plants are obliged to apply APCDs, while the old plants have undergone numerous improvements to control their emissions. Therefore, the average mercury emission factor for the coal-fired plants was dramatically reduced.

Conclusions

This study developed the measurement of mercury in flue gas based on aluminum matrix sorbent. A sorbent trap was used to collect gas phase Hg from flue gas and the Hg on trap was measured afterwards using CVAAS followed by acid leaching. Studies on Hg emissions and removals by existing pollution control devices in six different coal-fired power plants equipped with cold-side ESP and wet FGD systems have been carried out.

Hg adsorbed by sorbent was recovered completely by the leaching process. The average bias of the analytical technique was 2.68%, and the average recovery of spiked Hg^0 in tube 3 varied from 89.7% to 105% (average of 97%). Hg breakthrough in tube 2 was negligible compared to Hg content in tube 1. The average recovery of tube 1 and tube 2 was considered acceptable with a variation of 83.0% to 94.3% (average of 87.7%).

The mercury mass balances of the six power plants were 91.6%, 75.4%, 77.1%, 74.3%, 118% and 85.8%, respectively. Total Hg concentrations in stack gas ranged from 1.56—8.47 $\mu g/m^3$.

The relative distributions/fraction of Hg in bottom ash, ESP, FGD and stack ranged from 0% to 2.50%, 0% to 23.4%, 2.21% to 87.1% and 21.8% to 72.7%, respectively. The distribution profile varies with the coal type and the operation conditions of the plants.

ESP+FGD systems have the capability to remove Hg from flue gas. The average mercury removal efficiency of ESP+FGD systems is 44.2%.

The average emission factor of the six power plants in this study was 0.071 g/ton, which is much lower than the results obtained ten years ago in China.

References

40 CFR to Part 75. *Appendix K*: Quality assurance and operating procedures for sorbent trap monitoring systems, 2007.

ASTM D6784-02. *Standard test method for elemental, oxidized, particle-bound, and total mercury in flue gas generated from coal-fired stationary sources* (Ontario-Hydro Method), 2002.

BROWN, T. D. et al. Mercury measurement and its control: What we know, have learned, and need to further investigate. *Journal of the Air & Waste Management Association*, 49(6), p.628-40, 1999.

COOPER, J. A. Recent advances in sampling and analysis of coal-fired power plant emissions for air toxics compounds. *Fuel Process. Technol.*, 39, p.251-8, 1994.

DU, M. F.; ZHANG, Z. X. A study on the combustion characteristics of typical Chinese anthracite coals. *Journal of Engineering for Thermal Energy and Power*, 9(6), p.336-40, 1994.

GHORISHI, S. B. et al. Effects of fly ash transition metal content and flue gas HCl/SO_2 ratio on mercury speciation in waste combustion. *Environmental Engineering Science*, 22(2), p.221-31, 2005.

GUO, X.; ZHENG, C.; XU, M. Characterization of mercury emissions from a coal-fired power plant. *Energy & Fuels*, 21(2), p.898-902, 2007.

HALL, B.; SCHAGER, P.; LINDQVIST, O. Chemical reactions of mercury in combustion flue gases. *Water, Air, & Soil Pollution*, 56(4), p.3-14, 1991.

JIANG, J. et al. Development of mercury emission inventory from coal combustion in China. *Environmental Science*, 26(2), p.34-9, 2005.

KOUKOUZAS, N. K. et al. Mineralogy and geochemistry of Greek and Chinese coal fly ash. *Fuel*, 85(16): p.2301-9, 2006.

LEE, S. J. et al. Speciation and mass distribution of mercury in a bituminous coal-fired power plant. *Atmos. Environ.*, 40, p.2215-24, 2006.

MEI, Z. J. et al. Removing and recovering gas-phase elemental mercury by $Cu_xCo_{3-x}O_4$ ($0.75<x<2.25$) in the presence of sulphur compounds. *Chemosphere*, 70, p.1399-404, 2008a.

MEI, Z. J. et al. The effect of N-doping and halide-doping on the activity of $CuCoO_4$ for the oxidation of elemental mercury. *Appl. Catal. B-Environ.*, 78, p.112-9, 2008b.

MEIJ, R. The fate of mercury in coal-fired power plants and the influence of wet flue-gas desulphurization. *Water Air Soil Poll*, 56, p.21-33, 1991.

MEIJ, R.; SPOELSTRA, H.; WAARD, F. J. The determination of gaseous inorganic trace compounds in flue gases from coal fired power plants. *Proceedings of the 8th World Clean Air Congress*. Den Haag, Netherlands, 1989.

MEIJ, R.; VREDENDREGT, L. H. J.; WINKEL, H. T. The fate and behavior of mercury in coal-fired power plants. *Journal of the Air & Waste Management Association*, 52(8), p.912-7, 2002.

MUKHERJEE, A. B. et al. Mercury flow via coal and coal utilization by-products: A global perspective. *Resources Conservation & Recycling*, 52(4), p.571-91, 2008.

OLMEZ, I.; ARAS, N. K. Mercury determination in environmental materials by instrumental neutron activation analysis. *Trans. Am. Nucl. Soc.*, 68, p.171-3, 1993.

OTERO-REY, J. R. et al. As, Hg and Se flue gas sampling in a coal-fired power plant and their fate during coal combustion. *Environmental Science & Technology*, 37(22), p.5262-7, 2003.

PACYNA, E. G. et al., Global anthropogenic mercury emission inventory for 2000. *Atmos. Environ.*, 40, p.4048-63, 2006.

PAVLISH, J. H. et al. Status review of mercury control options for coal-fired power plants. *Fuel Processing Technology*, 82(2-3), p.89-165, 2003.

PRESTBO, E. S.; BLOOM, N. S. Mercury speciation in coal combustion flue gas: methodology, intercomparison, artifacts, and atmospheric implications. *Water Air Soil Poll.*, 80, p.145-58, 1995.

REN, D. Y. et al. Geochemistry of Trace Elements in Coals. *The Science Press*, Beijing, p.268-79, 2006.

RICHARDSON, C. et al. Effect of NO_x control processes on mercury speciation in utility flue gas. *Journal of the Air & Waste Management Association*, 52(8), p.941-7, 2002.

SHENDIKAR, A. D.; DAMPLE, A.; GUTKNECHT, W. F. *Report of research triangle institute*. No. EPA-600/7-84-089, 1984.

STREETS, D. G. et al. Anthropogenic mercury emissions in China. *Atmospheric Environment*, 39(40), p.7789-806, 2005.

UNEP. *Global Atmospheric Mercury Assessment*: Sources, Emissions and Transpor. 25th session of the Governing Council/Global Ministerial Environment Forum, Nairobi, 16-20 February, 2009.

USEPA – U. S. Environmental Protection Agency. *Method 30B*: Determination of total vapor phase mercury emissions from coal-fired combustion sources using carbon sorbent traps, 2007.

USEPA – U. S. Environmental Protection Agency. *Characterization and management of residues from coal-fired power plants*, 2002b.

USEPA – U. S. Environmental Protection Agency. *Control of mercury emissions from coal-fired electric utility boilers*. EPA-600/R-01-109, 2002a.

USEPA – U. S. Environmental Protection Agency. *Mercury Information Collection Request* (ICR), Part III, 1999.

USEPA – U. S. Environmental Protection Agency. *Method 29*: Methodology for the determination of metals emissions in exhaust gases from hazardous waste incineration and similar combustion processes, 1992.

USEPA – U. S. Environmental Protection Agency. *Method 101A*: Determination of particulate and gaseous mercury emissions from sewage sludge incinerators, 1982.

WANG, Q. C. et al. Study on the contents and distribution laws of trace elements in coal in northeast China and eastern Inner Mongolia. *Environmental Chemistry*, 15(1), p.27-35, 1996.

YOKOYAMA, T. et al. Mercury emissions from a coal-fired power plant in Japan. *Sci. Total Environ.*, 25, p.97-103, 2000.

ZHANG, J. Y. et al. Mercury in coal and its effect on environment. *Advances in Environmental Science*, 7(3), p.100-4, 1999.

ZHANG, L. et al. Mercury emissions from six coal-fired power plants in China. *Fuel Process. Technol.*, 89, p.1033-40, 2008.

ZHAO, X. G. et al. Effect of SO_2 on catalytic oxidation of NO over Co_3O_4/Al_2O_3. *Chinese J. Catal.*, 21, p.239-42, 2000.

ZHENG, L. G.; LIU, G. J.; CHOU, C. L. The distribution, occurrence and environmental effect of mercury in Chinese coals. *Science of the Total Environment*, 384(1-3), p.374-83, 2007.

ZHUANG, Y. et al. Impacts of acid gases on mercury oxidation across SCR catalyst. *Fuel Process. Technol.*, 88, p.929-34, 2007.

17

Integrated watershed management for sustainable land use in Brazil: Soil and Water Resources

Teresa Cristina Tarlé Pissarra[1]

Introduction

Earth is a planet where a group of organisms lives and survives by interacting individuals to the mineral resources and the products of the biosphere that supports the activities developed within. We are in constant evolution. The environment is modified by natural processes and human activities. Thus, we have to deal with land use and its cover relationships!

The concept of sustainability has evolved to meet changing needs and realities, technological developments and new understandings of ecosystem functions. According to White (2013), sustainability means different things to different people and remains an elusive concept, but it does provide a starting point for conversation between individuals from different disciplines, occupations and interests.

1 College of Agricultural and Veterinarian Sciences, UNESP – Sao Paulo State University, Rural Engineering Department, Jaboticabal, Sao Paulo, Brazil. E-mail: *teresap@fcav.unesp.br*.

The question is: How can we live in a better place in a sustainable way?

The environment is modified by natural processes and human activities. This is not simple and deals with land use and its relationships (Campos et al., 2012; Pissarra et al., 2009).

Human history in Brazil began with the arrival of the first indigenous people, thousands of years ago by crossing the Bering land bridge into Alaska and then entering the rest of North, Central and South America. The country's continental dimensions, contrasting climates, population distribution, economic and social development, among other factors, resulted in a wide array of differences ways of life between its geographical regions. NOWDAYS: How are we going to look to the future?

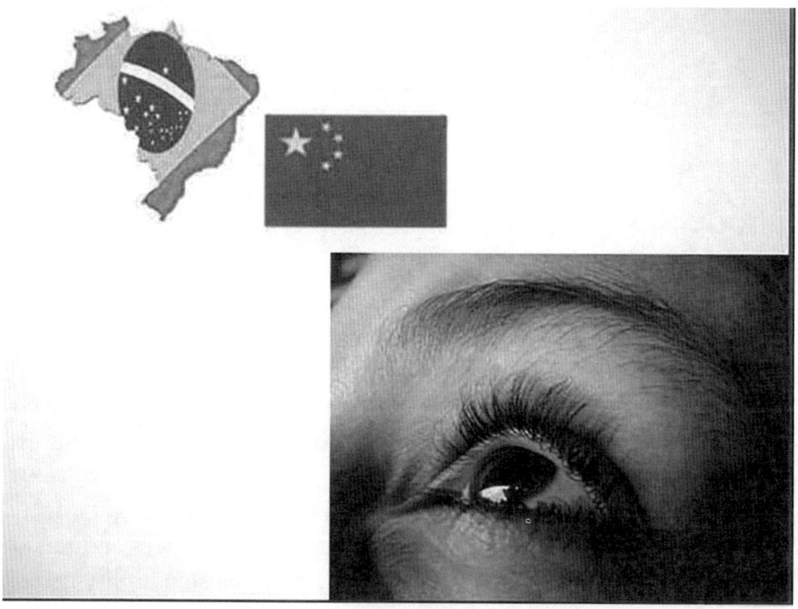

Integrated Watershed Management for Sustainable Land Use

Human development impacts the natural environment and transforms the land in different covers and uses (Figure 1).

Figure 1. Transformation of the landscape from its natural ecosystems.

The processes of land use by human activities determine land cover and modify our environment (Marzluff; Hamel, 2001; Kerr; Cihlar, 2005) (Figure 1), and we need to analyze the landscape and understand the support of the proprieties on its conversion to improve our settlement (Figure 2).

Figure 2. The landscape analyses to understand the proprieties' support.

The effect of land use on biodiversity is strong and information from the past can be used to develop solutions for natural resource management, including those regarding water, soil, nutrients, plants and animals (He et al., 2000; Pissarra, 2011).

Land use is the human use of land and involves the management and modification of natural environment or wilderness into modified environments such as fields, pastures, and settlements. It has also been defined as "the arrangements, activities and inputs people undertake in a certain land cover type to produce, change or maintain it" (Fao, 1997; Kutter; Neely, 1999; Marzluff; Hamel, 2001).

Watershed

How to modify the land?

First, we need to understand the concept of watershed management (Tomer, 2005; Pissarra et al., 2009). What is a watershed? It is a drainage basin or rather the area of land that contributes water to a particular watercourse (Figure 3).

Figure 3. Watershed.

A watershed consists of the topographic structures throughout a landscape (*land area*) from which water infiltrates and drains into a particular stream, river, or lake (Pissarra et al, 2009). The watershed area is related to the interdependencies between upstream, downstream and land use, and is defined as a set of resource-management practices that are planned and implemented to provide a sufficient source of quality soil and water to sustain human society and natural ecosystems (He et al., 2000; Tomer, 2005).

The status of drainage basin geomorphology forms the streams' orders (Horton, 1945; Strahler, 1957, Pissarra et al., 2010) that are correlated to the drainage area, but this is also regionally controlled by topography and geology. The watershed is made of many parts that cause the processes to modify the natural ecosystems over the years (Figures 3, 4).

Figure 4. Watershed and its processes (USGS, 2012).

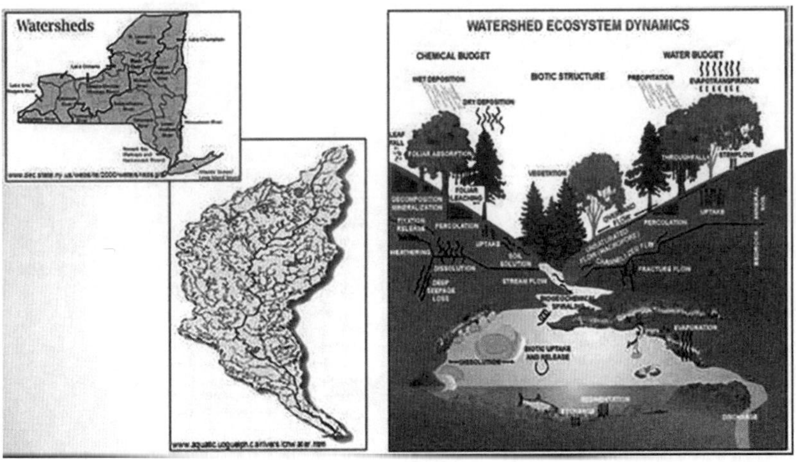

The drainage divide is the set of high points that separate one watershed from another. The main stream of river is the primary downstream segment of a river and the tributaries are the streams that flow into a main stream (or parent) river channel networks, in a particular lithology or similar geological terrain, in a random process. The spatial structure of channel networks was described by Horton (1945), who initially developed one-dimensional laws of drainage composition in the form of ratios, which is followed by many researchers. The outlet is the particular point where the water drains. The valley is the low area of land between hills or mountains, typically with a river or stream flowing through it. The upland is the area of high or hilly land. The riparian zone is the interface between land and a river or stream. The upland zone has a high sediment supply and low storage. The illuvial transport zone transports sediment, which decreases downstream.

We cannot think of a watershed if we do not think about the hydrologic cycle and a certain amount of soil that sustains species biodiversity.

The hydrologic cycle describes the continuous cycling of water from the atmosphere to earth to the oceans and back again (Figures 4, 5).

Figure 5. The hydrologic cycle/Water on the Web: developed by Merric (2003).

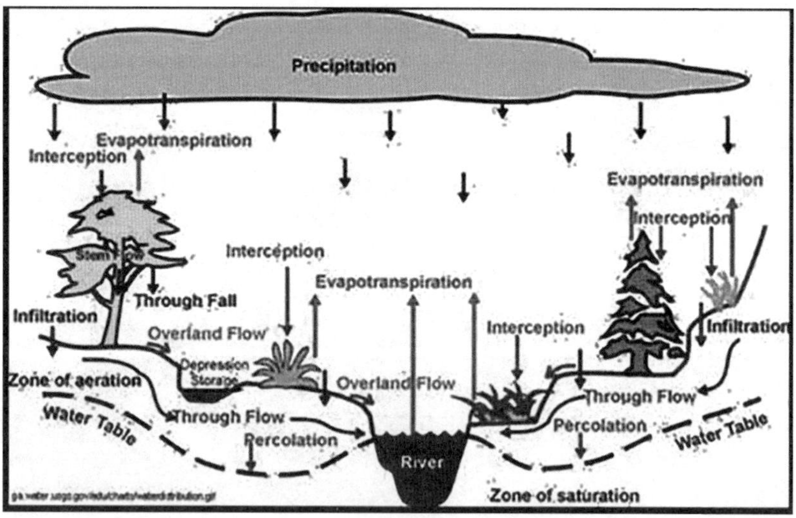

The process of evaporation, vertical and horizontal transport of vapor, condensation, precipitation, and the flow of water from continents to oceans are the major issues in determining climate changes and they influence surface vegetation and clouds, snow and ice formation and soil moisture (EPA, 2013). The hydrologic cycle is responsible for 25 to 30 percent of the mid-latitudes' heat transport from equatorial to polar regions (EPA, 2013; USGS, 2012).

Groundwater is the rain that falls then seeps through the soil down to the water table. The U.S. Geological Survey considers that ground and surface water interact throughout all landscapes from the mountains to the oceans. The surrounding upland change is a complex process that occurs from modifications in land cover to the land conversion process in conjunction with concerns regarding sustainability (USGS, 2012; Winter, 1998).

The amount of soil is from the processes of its genesis, operating under the influence of environmental factors that sustains the biodiversity of the species. We must think about soil complex to give us a logical framework for understanding the relationships between particular soils, the landscapes and the ecosystems in which they function (ANA, 2012; USGS, 2012).

Watershed management for sustainable land use

How to modify land use?

Evaluating the environmental suitability in a watershed area is a major concern for analyses of each field and is included as part of sustainable development. Each watershed is unique and includes particular elements, such as land uses, and requires individual attention (Tomer, 2005; Pissarra et al., 2009; Qi; Altinakar, 2011). It begins with understanding the basic concepts of watershed functions and values and learning how to protect or maintain a desired condition of that water resource (Tomer, 2005; Loveland; Bellamy, 2005).

The natural processes and its human-induced impacts modify the environment (Figure 6).

To develop a better watershed management scheme for sustainable land use we firstly need to observe the landscape and have a multidisciplinary discussion about land use and its consequences. This phase is called problem definition and will be involved in the solving process (Pissarra et al., 2009; Rocha; Kurtz, 2001; Rocha, 1997).

The development of watershed management strategies involves the Nation, the State and the Counties' objectives and visions, in regards to socio-economic goals and environmental laws.

The Brazilian Hydrographic Division consists of 12 main watersheds – Hydrologic Regions: 1. Amazonica; 2. Tocantins-Araguaia; 3. Sao Francisco; 4. Atlantico NE Ocidental; 5. Atlantico NE Oriental; 6. Parnaiba; 7.Atlantico Leste; 8. Atlantico SE; 9. Parana; 10. Paraguai, 11. Uruguai; 12. Atlantico Sul (ANA, 2012) (Figure 7).

Figure 6. Observing the landscape.

The National Water Agency's (ANA)/Brazil mission is to implement and coordinate joint and integrated watershed management resources and provide regular access to water, promoting its sustainable use for the benefit of current and future generations. The main objective is to respect the social, economic and environmental diversity with commitment, transparency, technical expertise, proactivity and public spirit, in order to be recognized by society as a reference for managing and regulating water resources and promoting sustainable water use (ANA, 2012).

The methodologies for land use planning are supposed to follow the potential land use practices and watershed management (ANA,

2012; Pissarra et al., 2009; Tomer, 2005). The data collection and evaluation are developed in a multidisciplinary way to improve the decision-making options. The hydrologic regions, land use, soil, water and the knowledge about the ecosystems functions and best management practices must be considered (Girolamo; Porto, 2012; Qi; Altinakar, 2011; Huang et al., 2012; White, 2013).

Figure 7. Brazilian Hydrographic Division (ANA, 2012).

As part of the *Sao Paulo State Land Use Framework* in Brazil, a Research Group at UNESP is studying an Integrated Watershed Management scheme for sustainable land use in Corrego Rico's Watershed (Figure 8).

Watershed management for sustainable land use will improve soil and water quality. Land uses practices impact hydrologic processes and the research group is supposed to identify ways to manage these impacts across a mosaic of soil and topography. The group is composed by many organizations, such as: *CATI-Coordenadoria de Assistência Técnica Integral; UNESP-Universidade Estadual Paulista/FCAV-Jaboticabal; SAAEJ-Serviço Autônomo de Água e Esgoto; CETESB-Companhia de Tecnologia e Saneamento Ambiental; CDA-Coordenadoria de Defesa Agropecuária da Secretaria de Agricultura e Abastecimento do Estado de São Paulo; PMA- Polícia Militar Ambiental do Estado de São Paulo/Regional jaboticabal; Prefeituras Municipais de Jaboticabal, Monte Alto, Taquaritinga, Santa Ernestina e Guariba; COPLANA-Cooperativa de Plantadores de Cana; FATEC-Faculdade de Tecnologia de São Paulo* and some others members and environmental organizations.

Figure 8. Corrego Rico's Watershed.

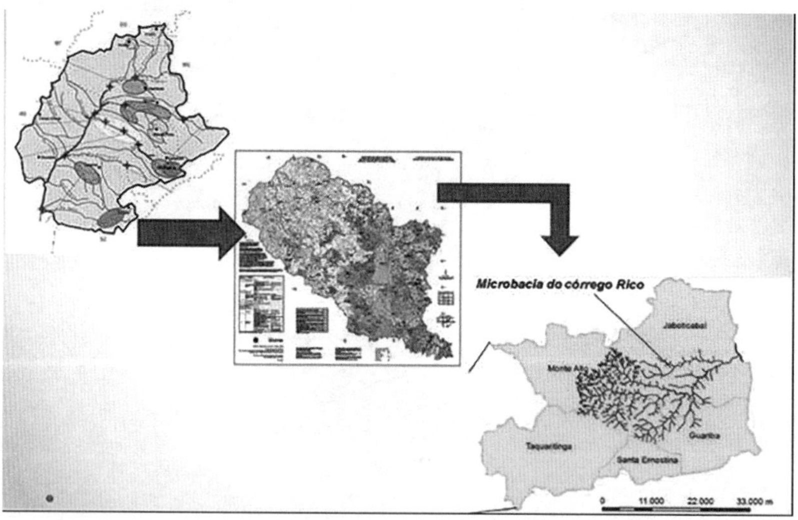

The main problem for defining Corrego Rico's watershed were the cropping systems' impact on the watershed hydrology due to soil disturbance and influence on the soil's capacity to support plant growth; the land management practice that alters soil conditions and the soil community of micro-, meso- and macro-organisms; the erosion processes; the population increase in urban areas, the land uses without soil conservation; and water quality. Some studies conducted in Brazil and China suggested similar problems and present some solutions, such as those by Furlani et al. (2008); Rodrigues; Pissarra; Campos (2009); Valle Junior et al. (2010); Fang et al. (2012); Silva et al. (2013); Huang et al. (2013), confirming that there are common research projects in both Brazil and China.

The soil survey of existing conservation systems and the tree species of some forested areas; the riparian areas and the physical characteristics of the watershed were mapped in a geographic information system (Nazário, 2009).

The group identified the alternatives and had to adopt some management strategies, such as tillage, crop rotations and use of plant residues and manure, changes in soil habitats and the food chain, and alterations in soil quality or the capacity of the soil to perform its functions. Farming practices that minimise soil disturbance (ploughing) and return plant waste to the soil, such as no-tillage farming and crop rotation, was applied at the farmers' areas and allowed a slow process of rebuilding the soil's original features and restoring the soil's organic matter. Reducing tillage tends to result in increased growth of fungi, including mycorrhizal fungi (Furlani et al., 2008; Silva et al., 2013).

Forest is a very interesting land cover, allowing water to enter the ground and recharge the water table. In Brazil, the riparian areas must be preserved along the rivers and its management offers the opportunity to improve the stream flow regimen and water quality, bringing many benefits to natural ecosystems (Pissarra et al., 2009).

The group developed a Program that selected some "springs" and lands along the drainage net to be recomposed. The strategies to be followed to achieve preservation in these areas included: assisting the farmers to plant tree species, encouraging forestry management and environmental education. The replanted area will protect a significant number of animals' habitats, wetlands and ground water recharge areas. The program is called "*Reflorestando as nascentes*" (Reforesting the Spring and Riparian Areas). Usually, children from Public and Private Schools go to farmers lands to help them plant the trees (Figure 9).

At urban canters, paved areas make the water flow faster into streams. The group studied the park areas to rebuild a new concern about urban reforestation.

Figure 9. Environmental Education at Riparian Areas.

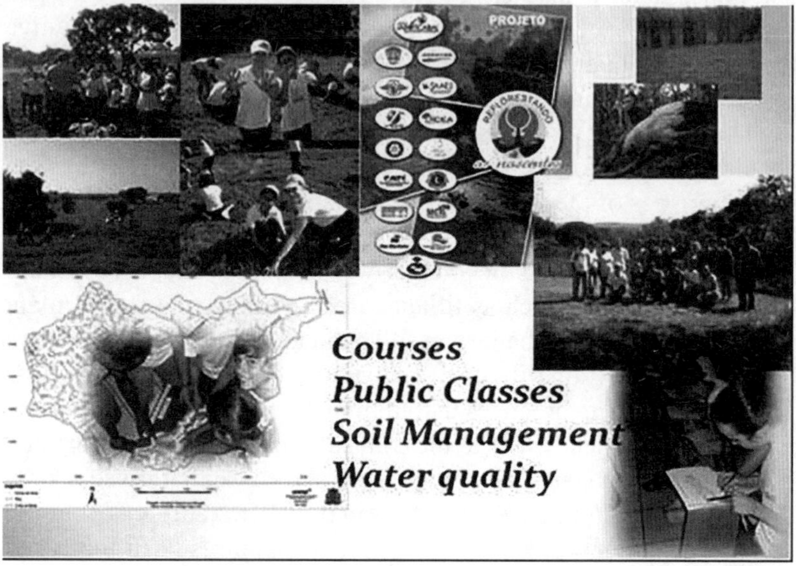

Final Considerations

Technical experts provide many benefits and an effective watershed planning depends on an objective assessment.

Land uses affect key hydrologic processes such as runoff, infiltration, evapotranspiration, percolation etc. The urban industry and farming communities need to develop a new concern regarding land management issues such as improving water and soil qualities and water availability to plants, ensuring better access to sources of fuel, controlling the soil's erosion and land degradation, and in particular avoiding soil nutrient depletion and pollution of the air, soil and water resources. It is important to identify water quality problems associated with land management activities (Tomer, 2005; Rodrigues et al., 2009) so erosion control at watersheds will minimize the pollutants (Tomer, 2005; Huang et al., 2013; Yu et al., 2013; Shi et al., 2012; Zhu; Schmidt; Bryant, 2012).

A number of computer tools can help perform a watershed assessment. Using a geographic information systems (GIS) database provides ways to view and evaluate combined map data that includes soil surveys, land use and terrain analyses and remote sensing (Pissarra et al., 2013; Girolamo; Porto, 2012; Ker; Cihlar, 2005).

In order to improve knowledge, it is necessary to teach in an Environmentally Educational way. Environmental courses, public classes for teaching about soil management and water quality in watersheds should be provided at schools. Learning better ways to live on Earth will make conserving ecosystems for future generations possible.

Acknowledgements

With thanks to TOP CHINA Program from Santander Universidades;

To Shangai Jiao Tong University, China;

To UNESP/FCAV/Rural Engineering Department;

Special thanks to the research team for their instructive comments and suggestions regarding the manuscript; and to all my research colleagues who work at the Corrego Rico's Watershed. The financial support at some projects at Corrego Rico Watershed was provided by UNESP/FCAV/PROEX; CNPq; CAPES, FAPESP and the São Paulo State Government.

References

ANA 2012. AGÊNCIA NACIONAL DE ÁGUAS (Brasil). Available at: http://ana.gov.br/. Access in: June 8th, 2012.

CAMPOS, S. et al. Geotecnologias aplicadas na evolução do uso da terra na microbacia do córrego Capivari, Botucatu (SP). *Revista Científica de Engenharia Florestal*, 19(1), p.1-8, 2012.

EPA 2013. United States Environmental Protection Agency. Available at: http://www.epa.gov.

FANG, N. F. et al. The effects of rainfall regimes and land use changes on runoff and soil loss in a small mountainous watershed. *Catena*, 99, p.1-8, 2012.

FAO. Food and Agriculture Organization of the United Nations. *Land Cover Classification*. Environment and Natural Resources Services (SDRN). 76p. 1997.

FURLANI, C. E. A. et al. Temperatura do solo em função do preparo do solo e do manejo da cobertura de inverno. *Revista Brasileira de Ciência do Solo* (Impresso), 32, p.375-80, 2008.

GIROLAMO, A. M.; PORTO, A. Land use scenario development as a tool for watershed managment within the Rio Mannu Basin. *Land Use Policy*, 29, p.691-701, 2012.

HE, C. et al. A conceptual framework for integrating hydrological and biological indicators into watershed management. *Lanscape and Urban Planning*, 49, p.25-34, 2000.

HORTON, R. E. Erosional developments of streams and their drainage basins: hydrophysical approach to quantitative morphology. *Geological Society of America, Bulletin*, v.56, p.275-370, 1945.

HUANG, J. et al. Watershed-scale evaluation for land-based non-point source nutrients management in the Bohai Sea Bay, China. *Ocean & Coastal Management*, 71, p.314-25, 2013.

HUANG, J. et al. Use of intensity analysis to link patterns with processes of land change from 1986 to 2007 in a coastal watershed of southeast, China. *Applied Geography*, 34, p.371-84, 2012.

KERR, J. T.; CIHLAR, J. Land Use Mapping. *Encyclopedia of Social Measurement*, 2, p.441-9, 2005.

KUTTER, A.; NEELY, C. L. (Ed.) The Future of Our Land: facing the challenge. Natural Resources Management and Environment Department. Guideline for integrated planning for sustainable management of land resources. Roma: FAO/UNEP: United Nations Environment Programme, 1999.

LOVELAND, P. J.; BELLAMY, P. H. *Environmental Monitoring*. Silsoe, UK: Cranfield University, 2005. p.441-8.

MARZLUFF, J. M.; HAMEL, N. Land Use Issues. *Encyclopedia of Biodiversity*, 3, p.659-73, 2001.

NAZÁRIO, D. J. F.; PISSARRA, T. C. T. *Levantamento de espécies arbóreas potenciais para sistemas agroflorestais, na formulação de trilha de interpretação e educação ambiental*. Trabalho de Graduação em Agronomia, Jaboticabal: UNESP/FCAV, 2009. 71p.

PISSARRA, T. C. T. et al. Caracterização hidrica da microbacia hidrografica do Córrego da Fazenda da Glória, Taquaritinga-SP. *Revista SODEBRAS*, v.6, p.1-5, 2011.

PISSARRA, T. C. T. et al. Morfometria de microbacias hidrográficas do córrego Rico, afluente do Rio Mogi-Guaçú, Estado de

São Paulo, Brasil. *Revista Árvore* (Impresso), v.34, p.669-76, 2010.

PISSARRA, T. C. T. et al. Informações básicas para o planejamento ambiental: Município de Jaboticabal/Secretaria Municipal de Agricultura, Abastecimento e Meio Ambiente de Jaboticabal. Jaboticabal: Funep, 2009. 70p.

QI, H.; ALTINAKAR, M. S. A conceptual framework of agricultural land use planning with BMP for integrated watershed management. *Journal of Environmental Management*, 92, p.149-55, 2011.

ROCHA, J. S. M. *Manual de Projetos Ambientais*. Universidade Federal de Santa Maria, 1997. 423p.

ROCHA, J. S. M.; KURTZ, S. M. J. M. *Manual de manejo integrado de bacias hidrográficas*. 4.ed. Santa Maria: CCR/UFSM, 2001. 302p.

RODRIGUES, F. M.; PISSARRA, T. C. T.; CAMPOS, S. Condições hidrológicas de uma bacia hidrográfica com diferentes usos do solo na região de Taquaritinga, SP. *Irriga* (UNESP Botucatu), v.14, p.158-69, 2009.

SHI, Z. H. et al. Modeling the impacts of integrated small watershed management on soil erosion and sediment delivery: A case study in the Three Gorges Area, China. *Journal of Hydrology*, 438-439, 156-167, 2012.

SILVA, R. P. et al. Qualidade da colheita mecanizada de feijão (Phaseolus vulgaris) em dois sistemas de preparo do solo. *Revista Ciência Agronômica* (UFC. Imp.), 44, p.61-9, 2013.

STRAHLER, A. N. Quantitative analysis of watershed geomorphology. *Trans. Am. Geophys. Un.*, New Haven, 38, p.913-20, 1957.

TOMER, M. D. Watershed Management. *USDA Agricultural Research Services, Ames, IA, USA*, p.306-15, 2005.

USGS 2012. U.S. Department of the Interior | U.S. Geological Survey | DOI Inspector General URL: http://www.usgs.gov/default.asp Page Contact Information: Ask USGS Page Last Modified: Wednesday, Dezembro, 2012.

VALLE JUNIOR, R. F. et al. Diagnóstico das áreas de preservação permanente na bacia hidrográfica do Rio Tijuco, Ituiutaba-MG, utilizando tecnologia SIG. Engenharia. *Agrícola*, 30 (3), p.495-503, 2010.

WHITE, M. A. Sustainability: I know it when I see it. *Ecological Economics*, 86, p.213-7, 2013.

WINTER, T. C. 1998. Ground water and surface water: a single resource. U.S. Geological Survey circular: 1139. II. Series. GB661.2.G76.

YU, D. et al. Detecting land use-water quality relationships from the viewpoint of ecological restoration in an urban area. *Ecological Engineering*, 53, p.205-16, 2013.

ZHU, Q.; SCHMIDT, J. P.; BRYANT, R. B. Hot moments and hot spots of nutrient losses from a mixed land use watershed. *Journal of Hydrology*, 414-415, p.393-404, 2012.

18

Better City, Better Life – Minimum Sustainable ArchitectureMinimum Sustainable Architecture

Silvio Stefanini Sant'Anna[1]

Introduction

The following study shows an experience that helps us better understand the conceptual and ideological meanings of sustainability that apply to architecture. It is very important to understand this new paradigm so that we can learn more about it, but it is especially vital for an architect who has a logical system of thinking about procedures. To thoroughly understand the principles of sustainability one needs to live an experience. One needs to feel the importance of the environment on the body. To begin this experiment, we must look to history. Let us study the past to learn more about the 5000 years of Chinese culture, which is a culture ten times older than that of Brazil.

By analyzing the meanings of Chinese characters, one notices concepts of simplicity and sustainability that have been present for thousands of years in this written language.

1 Universidade Presbiteriana Mackenzie, São Paulo, São Paulo, Brazil. E-mail: *silvio.santanna@mackenzie.br.*

Figure 1. Wood is symbolized as a tree.

Figure 2. Rice symbolizes food.

Figure 3. Book symbolizes culture.

I learned from a Chinese carpenter in Brazil that if you cut a 100 year-old tree you have to make something with this wood that will survive more than 100 years. All the temples in Shanghai are built using a timber structure and they have been standing for the same amount of years as the wood that was used to build them; these temples represent the culture of Chinese people. This is an example of how the idea of sustainability was already a preoccupation in the Chinese millennial culture, many, many years before Brazil was even discovered.

The recent history of architecture provides us with other examples.

When architect Mies Van Der Rohe created the slogan **Less is More,** around 1930, he was trying to influence society on the necessity to build quickly, more rationally, and with less focus on detail, which he deemed fundamentally important, especially after the two World Wars. It was important to rebuild cities in order to recuperate the chaotic economic systems of the countries after the wars. This was the start of modernism in architecture.

Some years later, in 1970, architect Robert Venturi, as an antidote to the Mies slogan, created the **Less is a Bore** phrase. It is in a book entitled "*Learn from Las Vegas*". Las Vegas is considered by many to be the most outlandish city in the world, with all kinds of entertainment and attractions. Its existence is proof that values change. At the same time that societies need normal lifestyles, there is also the need to sometimes cross over the equilibrium line and venture into a wild and crazy style with different values. This is pluralism. And it proves how ambiguous our society is. Robert Venturi was proposing to join post modernism culture with postmodern architecture. In my opinion it makes sense but it did not work well.

Today, when we are trying to change the paradigm, the concept of ecology makes us take preservation into consideration. However, the question is how to preserve all those natural things, if the human being has in its own nature production as an interface, an interchange with the environment

That is why the concept of sustainability exists.

It is very important to continue producing while keeping the natural environment as untouched as possible.

Another architect created a new slogan for this idea.

Do More with Less.

I would like to discuss an experiment that designed and built a vacation house inside a preserved tropical forest. Everything was done within all the legal frameworks in Brazil.

I will show you some pictures and then propose an exercise.

Insertion in Nature

A question that is always present in our discussions is to how to turn an environment filled with trees, rocks, birds, squirrels, orchids, grasses, rivers, waterfalls, insects, and a wide range of fauna and flora, into an environment for humans with all the conveniences of cities without interfering with its rich homogeneous and balanced context. However, including streamlined housing within a natural habitat as described above can be highly significant in terms of exercise and understanding the environment.

To understand this, we as architects must try to investigate in depth what it means to leave the chaotic and stressful city life to settle into a dense forest near a crystal clear waterfall. One cannot just concentrate on the physical location of the place, one must also think about the sociological aspects of the small community that lives in the area, with their traditions and customs that are immortalized by generations of *caiçaras sertanejos* (local fishermen) who know the quiet paths and the tough routes of that region in depth. This population, often composed of simple folk, is so deeply rooted in the land that it makes them eternal land owners through a hereditary feeling that cannot be bought or sold. One can only live with it. For this reason, one must also respect the inclusion of a building there, whatever its purpose. It is necessary to understand these meanings and begin establishing a space that first and foremost respects the difference between the cosmopolitan spaces that we are accustomed to. Architecture is a key element of this process. Viewing the building process as something that will visually interfere in the harmonious landscape means respecting the way of life that exists there (Figure 4). Thus, for the proposed project, it is not simply enough to simply meet the needs of a client or even the construction techniques that apply to the financial conditions of the project. One must take all the surrounding issues that involve the construction into consideration, which should prioritize its solutions and influences within a holistic paradigm; which is the key to striking a balance between necessity and respectability.

Architecture that must meet the following principles: respect the environment and understand it through the soil, trees, birds, insects, plants, wind, water, hills, stones, sounds, shadows, lights, day, night, the changing temperatures, the smells, and every other aspect that belongs to the natural world. Respect the local construction technology and understanding its technique as a characteristic of the local labor force, while not belittling more appropriate technologies, material availability, access to the construction site, transportation, logistics of work supervision, transit of local manpower, payment of wages, technical knowledge of the structural system by the architect and the general contractor, and the understanding of all the technicalities involving the project. Respect the social, human, cultural and economic factors of the region, and all other aspects that might interfere with the general lines of the project.

Figure 4. Sketches by the Architect. In the first deployment of the house on the ground in the form of the indigenous tribe and the second in a detail view of the house off from the floor.

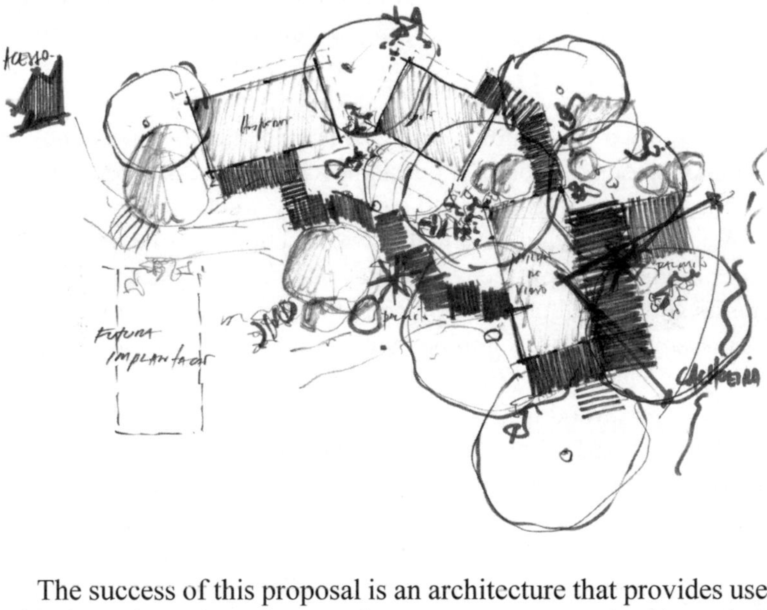

The success of this proposal is an architecture that provides users with the redemption of perception senses as a source of knowledge and discovery and brings satisfaction and a contemplative feeling of ecstasy that are often forgotten or even unknown. Given the immaculate nature of the site before construction and the purity of feelings that the project may still transmit to the landscape, we believe that this architecture can and should serve as for the pursuit of a professional work ethic that has always followed us in our studies and that is certainly a result of our work.

Building and Sustainability

The current tendency in architecture considers sustainability to be a fundamental reference for respecting the environment. The idea of preserving natural resources for future generations is the ultimate sign of respect that we have an obligation to exercise. Since the 70's, with the oil crisis, more attention was given to the extraction of natural resources, until we realized the importance of sustainability issues. Along with the broadening of preservationist ideas came the principle that sustainable construction requires designing qualities that have always been present in our training, in addition to the fundamental principles of architecture, which have been disregarded in recent years in light of a technology-oriented consumption and fads that try to reproduce an international aesthetic for buildings. Any citizen with common sense will not oppose the guidelines of sustainable construction. All that is needed is access to technology and knowledge of these guidelines. Usually what occurs is that the technology involved in sustainable construction is still expensive as it does not have government subsidies that promote the spread of its benefits and the cultural acceptance of the construction procedures. The dissemination of technology and tax incentives would help promote beneficial competition to the consumer, making its acceptance happen faster. Wood is a good example.

In small- and medium- sized constructions, wood could fulfill an important role in the characterization of the house. Since it is a natural material, easy to handle and tactile to man, wood can generate different landscape possibilities with new sensory and perceptual possibilities.

This method of construction is widely disseminated in the United States and is known as "steelframe" or "woodframe". It has a fundamental characteristic: the inversion of the concept in construction of the building, which is known as a simple assembly. Starting with a rationalization in the design projects, the dry construction promotes facilities that transform conventional construction into a system of assembly with specific criteria in the pieces that are put together.

There are no leftovers or waste. This concept is fully compatible with sustainability and dry building. The dismantling of the buildings allows 90% of the materials to be reused. The cultural factor in our society prevents fast diffusion of alternative building systems that are better suited to our needs.

Case Study

Unlike all of this, the project of the house "Vila Barulho D'Agua" was more than a professional exercise and became an experimental innovative laboratory. This process, with actions ranging from the understanding of programmatic requests to the uncertainties of unlimited freedom perpetuated with the development of the project, has provided an opportunity to exercise this investigative and distinct type of architecture. This exercise started with the characteristics of the region, since we were unable to do a "plan altimetry" study, due to the difficult location of the terrain, which was 300 km away from our office. Therefore, we had to include in the development of the house, design methods related to new materials and the construction of new paradigms, which created a differentiated execution of the work that what we were used to.

While we developed the project it was inevitable that we consider the forest and the impressive environmental condition, the small availability of funds, the personal needs of the client, the time required and the difficulty in monitoring the work, given its distance. However, it was interesting to solve real problems in the development and construction of a vacation home of approximately 120 m^2, located in an area of 2,100 m^2 of native vegetation with an abundance of rocks, bordered by a river with many waterfalls, 6km from the center of the city of Paraty, in the state of Rio de Janeiro. With this scenario we approached sustainable solutions, and intuitively solved fundamental issues so that the desired project would fulfill its social objective.

While visiting the site and admiring the glistening terrain covered by indescribable vegetation, dotted with rocks and stones of various sizes, touched or hidden by a dense undergrowth of plants, the first sensation was to give up the challenge and just contemplate the pure and melancholy nature of the environment. Past this first impression, and having taken in the magic of the area, our minds began to produce and imagine shapes and designs in an attempt to translate the dreamy encounter with nature. We would not overturn the beauty of the place as it was presented. As an alternative we used organic materials, consistent with the site, such as certified wood, processed by advanced technological systems and treated against insects and moisture. These actions, if worked properly and rationally would give the desired characteristics to the project, besides setting up an environmentally friendly project. In the first sketches, the clean forms, along with the alternative of making functional sectors of the house, defined the use of modules as a tendency, as a means to respect the small spaces left between the existing trees. The use of a single block proved to be impossible because of the presence of large trees around the land. Implementing these characteristics would require clearing the land, cutting down many trees, which was undesirable. The principle of preserving the forest always prevailed. This same procedure of leaving the land as untouched as possible was also adopted, as was reconstructing the vegetation that grew between the stones.

The solution to make smaller modules around the ground was joined by a proposal of non-interference in the topography, so that these modules minimally touched the ground, maintaining all its permeability. This solution was crucial to establish a comfortable thermal condition in the house, since the space between the ground and the floor of the building promotes an air passage that prevents humidity, which is very common in this region.

The defining aspect of the project were the modules and the high-ground construction, which helped us solve another concern of the client, which was to construct the house in stages. After defining the structure in wood and tiled roof, the next important step was to define the seals necessary to implement the conditions we wanted: speed, lightness, durability and strength. This search led us to OSB, which is an abbreviation for Oriented Strand Board, measuring 2.44m x 1.22m in various thicknesses, with the necessary characteristics for the intended use. The texture of the wood in chips was our particular challenge. Using the wood as a seal without any external coating seemed too daring. However, we believed that with an extended overhang, which would remove the contact of rain water, and with a suitable paint job, the seals would be properly made. The panel system, structured as a

Figure 5. Overview of the guest module. Image from the architect.

sandwich, together with the glass panels to ensure transparency, defined the synthesis and purity of design that we wanted. The modules that sit on higher ground, the wooden structure, the use of clay tiles on the roof plus the seals made with the OSB panels and glass, presented the definitions needed to design the details of this enriching project.

Figure 6. View of the glass module of the living room. Image from the architect.

Starting from the dimensions of the panels, the project had such a tight consistency that the measurements of the rooms came to be variations in the measurements of the panels. The biggest quality of this process is to use the maximum of materials and avoid waste. Due of the use of whole parts and because there were no bricks or concrete beams

Figure 7. Detailed view of the master bedroom window. Image from the architect.

and columns in the construction, the project established different procedures than those used in conventional buildings. Following a simplified drawing of the solutions, the details were established within the necessary proportions to create a harmonious balance between the living space and the use of the forest as a neighbor. The outside decks, used as passageways between the modules, are slightly loose from the ground, creating a subtle design among the rocks and the tropical vegetation of the central part of the terrain. The path that is determined by the floor provides interesting visual perspectives, likes photo frames of a beautiful landscape. The other decks provide the continuity into social areas, outside the glass pavilion, where the protective canopy of the trees create a shaded cover so one can observe the waterfall, birds and squirrels. Being able to relax in this area on top of pillows, while eating breakfast or enjoying a barbecue, are part of the purpose of this deck. This project certainly considers wood and technological trends in its production and application as key points to continue developing the idea of architectural prototypes consistent with our time, while never forgetting the triad:

References

OKAMOTO, J. *Percepção ambiental e comportamento*. Visão Holística da Percepção Ambiental na Arquitetura e Comunicação. São Paulo: Editora Plêiade, 1996.

RUANO, M. (Org.). *Um Vitruvio Ecológico*. Princípios y Práctica Del Proyecto arquitectónico sostenible. Barcelona: Gustavo Gili, 2007. p.59-79.

SANT'ANNA, S. S. *Vila Barulho D'água*. Um caso de Arquitetura Sustentável PRÓ-Livros. São Paulo, 2005.

19

The Sewage Treatment Plant with a Case Study in Cachoeiro de Itapemirim, ES, Brazil

Sandra Novaes Coelho[1]

Introduction

With the objective of facing the population growth challenge that urban cities around the world are facing, China raised the general theme of the program Better City, Better Life, which was the moto of a series of activities carried out in China, with the focus being to analyze urban plans, observe economical sustainability, cultural, social and environmental standards, based on local features.

The proposed theme should have two main focuses, which were: 1) Urban planning, and 2) Environmental Sustainability, which reflect the aspects that are at stake in the development of cities around the world.

Thus, it was decided to perform an analysis on the sewage treatment plant in Cachoeiro de Itapemirim, a city with about 200 thousand inhabitants, in the south of Espírito Santo, Brazil.

1 União Social Camiliana Cachoeiro de Itapemirim, Espírito Santo State, Brazil. E-mail: *sandranovaes@saocamilo-es.br*.

Sanitary System in Brazil

The World Health Organization defines a sewage plant as the control over men's physical environment that exerts or can exert deleterious effects over physical, social and mental well-being, that is, over a citizen's health.

However, it is only now, in the XXI century, that access to the sewage network is available to half of Brazil's population (50.9%), according to FGV/Treats Brazil.

The president of the PNUMA Brazil Institute warns that the resources for basic sewages plant need to increase if the country intends to reach the United Nations Milenium Developmnt Goals.

In order that this country can reach these goals, which were adopted by Brazil during the 2002 UN Summit on sustainable development in Johannesburg, it is important to increase the investment in sewage, which is the point of view of the PNUMA Brazil Institute, Haroldo Mattos de Lemos.

The PNUMA Brazil Institute is a nonprofit private organization whose main activity is publishing the results from studies conducted by the PNUMA-UN Environment Program, as well as promote participation in educational activities and environment awareness. Mr. Haroldo de Lemos adds, "The actions in the sewage area need to be understood as 'health prevention'".

Mr. de Lemos argues that "although the UN data show that one dollar invested in water supply and sewage plant means saving at least five dollars in hospitals, sewage is not yet a political priority in the country". And that "Oscips and NGOs need to service the area. They contribute by spreading of the need for more awareness towards sewage and for more support from the population in order to push politicians to define sewage as one of the nation's priorities".

A study carried out by the *Ministério das Cidades* (Ministry of the Cities) affirmed that, if the current trend goes on, Brazil will hardly decrease the percentage of people who need proper sewage attendance by half until 2015, as is forecast by the seventh Objectives for the Development of the Milenium.

The Trata Brasil Institute evaluated the sewage services in 81 of the biggest Brazilian cities, with over 300 inhabitants. The analysis was based on the *Sistema Nacional de Informações sobre Saneamento* (SNIS – National System of Information on Sewage), which was published in March by the Ministry of the Cities, which gathers information on the water supply and sewage service offered spontaneously by the companies that work in this area.

Between 2003 and 2008, there was a development of 6.9% in sewage collection and 11.5% in the treatment in those cities. However, 5.9 billion liters of non-treated sewage are still dumped into the environment. Therefore, the Trata Brasil concluded that universalizing the services will not be enough to reach the deadline of 2027.

In big cities, for example, like in Sao Paulo, it has been 15 years since they started the recovery of the Tietê River, which is the city's main river. Water treatment is very expensive, and the surreptitious sewage connections are one of the greatest problems.

Professor and engineer Mário Thadeu Leme de Barros, from the Escola Politecnica at USP – São Paulo University, is a master in hydro- and sanitary- engineering, and understands the problem very well: "All the rivers here in São Paulo are dead, due to the excessive dumping of sewage into their waters. And the main source of pollution of the river sewage is from domestic sewage," he informs.

This professor believes that the Tietê River suffers consequences from two other great problems: the surreptitious connections, leaking in the water and sewage tubes and street littering that goes into the river. It is the so called diffused pollution.

As one looks at the river from above, it seems that it is dead. "I am not so pessimistic. I believe that a few years from now we will see this river clean," said environmental engineer Mônica Porto, "if the efforts for the treatment go on as they have been for the last 15 years. I hope that in 15 or 20 years from now it will be much better than it is today," she pointed out.

According to the World Health Organization, each Real (Brazil's currency) invested in sanitary services is equivalent to four Reals saved in public health.

In summary, 75% of the sewage collected in the Brazilian cities is not treated (ONU/-3 and Instituto Brasileiro de Geografia e Estatística – IBGE/04).

Up until 2008, open sewers, known as "lixões" have been the final destination for wasteload allocation in 50.8% of Brazilian municipalities, but this figure has significantly changed over the last twenty years: in 1989, they represented the final destination of solid waste in 88.2% of the states. The Northeast (89.3%) registered the greater number of municipalities that had their waste sent to the wasteload allocations, while the South regions (15.8%), and the Southeast (18.7%) regions presented the lowest percentages.

Table 1. Solid Waste: Wasteload allocations are the final destination in half of the states in Brazil.

Final destination of the solid waste, by unit of the waste (%)			
Year	Open sewers	landfill	Sanitary landfill
1989	88,2	9,6	1,1
2000	72,3	22,3	17,3
2008	50,8	22,5	27,7

Parallel to this, there was an expansion in the destination of waste to landfills, which was a more adequate solution and went from 17.3% of the municipalities, in 2000, to 27.7%, in 2008.

In the whole country, nearly 26.8% of the municipalities that had a service to maneuver solid waste were aware of waste collectors, individuals that collect waste for their survival. Most of them were in the West and Northeast: 46% and 43%, highlighting municipalities in Mato Grosso do Sul (57.7% were aware of collectors), and Goiás (52.8%), in the Central-West and the Northeast, the municipalities of Pernambuco (67%), Alagoas (64%), and Ceará (60%).

Sanitary System in Cachoeiro de Itapemirim

Cachoeiro de Itapemirim lies in the south of Espírito Santo State and is covered by the Itapemirim River, which is the main water supply source for eighteen municipalities, and which has seven micro basins. Nowadays, the population is aware of the fact that it is important to protect the river basin, but it is still paramount that the preservation work must be effective and continuous in order to ensure a quality of life for future generations. In addition, visual and economic reasons are just as important. After all, the river crosses the whole town. In the cities of Marataízes, Piúma, Jerônimo Monteiro, Presidente Kennedy, Mimoso do Sul, Muqui, Alegre, Rio Novo do Sul, which surround Cachoeiro de Itapemirim, in Espírito Santo State, waste load allocation is still adopted. However, the latter is now a national reference in terms of sewage treatment, as it is one of the cities in the State that has been awarded the gold rank according to the Espirito Santo 2010 Quality Award, and currently, it is aspiring to the platinum rank.

Scenario up to 1998

Water supply: Before the Foz do Brasil started its activities in Cachoeiro de Itapemirim, there was constant lack of water supply. The services were performed by a municipal autarchy. In order to properly solve the problem, it was necessary to maneuver the system in its several areas, with at least the minimum amount of water.

Water Leakages: The supply system used to lose 56% of the water distributed.

The leakages, according to top employees, would go up to 2,000 leakages per month.

Equipment and Old Technologies: The employees used to take the city bus to work to maintain t water and sewage. There used to be no equipment, tools or adequate vehicles. Both the equipment and technologies were quite old, as the maintenance net and the management were inadequate.

Hydrometer: Only 76% of the water used was measured. The hydrometers were quite old and re-used, which compromised the reliability of the measurements taken.

Consumption: The average consumption of water in the municipality was 190.60 liters per inhabitant.

Sewage treatment: Sewage treatment was a distant reality, considering the problems faced by the water supply. Twenty one million liters of sewage were dumped into the Itapemirim River on a daily basis.

Water quality risks: The lack of infrastructure in the laboratory and the old equipment harmed the quality of the water. There was no quality standard guarantee, even if it fulfilled the minimum required by the Ministry of the Cities.

History of the Foz do Brasil

In 1998, as the municipality opened a bid for the water supply service in the city of Cachoeiro de Itapemirim, Citágua Águas de Cachoeiro S.A., won the right for its water supply management.

Having had a loan from BNDES – *Banco Nacional de Desenvolvimento Econômico e Social*, provided in 2002, the municipality could have its water and sewage treatment plants built. The investment program, implanted by Citagua, covered the rehabilitation and widening of the water supply system, complementation of the collecting area and beginning of sewage treatment, training its employees and administrative staff and updating the group through computerized activities and implementing a quality program.

After ten years' concession, in July 2008, Citagua passed their work onto Odebrecht Engenharia Ambiental (OEA), from the Odebrecht. Organization. In 2009, the name changed to Foz do Brasil, and then the company continued the sewage treatment project, within a process that involved a concession agreement with the municipality. The terms say that the Foz do Brasil, despite being a private institution and administrator of water supply and sewage services, the estate belongs to the municipality, and the services are controlled by *Agersa – Agência Municipal de Regulação dos Serviços Públicos Delegados de Cachoeiro de Itapemirim.*

The population of Cachoeiro de Itapemirim is about 200 thousand people, the concession at this time covers 88% of the sewage produced, and the Foz do Brasil distributes quality water to 100% of the population. Their aim is to treat 90% of the sewage volume in this town by 2012. 12 km of interceptors, 10 km of trunk collectors and 88 km of sewage collecting networks have been built, in addition to the Sewage Treatment Station, with a capacity of 560 liters per second.

The Foz do Brasil Sewage Plant

Basically, a sewage treatment station aims to treat domestic sewage, in order to enable it to be cleanly dumped into rivers, streams, lakes or the ocean.

The residual domestic waters, or sewage, are directed by the home collectors toward a collecting net called a rough sewage emissary, which passes through the city streets. These interconnected home pipes go through interceptors up to the green pipes which go all the way down the Itapemirim River, and then the sewage flows by gravity into the sewage treatment plant. Such collecting nets converge through the interceptors (with wide dimensions) up to the bombing station or directly to the sewage treatment station, depending on the local topography.

Biofilter and Sand Box: The station receives the effluent through a sanded system, where the pre-treatment happens, which is made up of scrapers, a sand box and transporting spirals, which are responsible for trapping thick objects in the effluent while the gases collected are sent to the Biofilter, which is also efficient at absorbing the odors. With the retention of such solids, the effluent follows the entrance channel into the aerial tanks, according to the Unitank system.

Therefore, at the Sewage Station Plant the rough impurity is initially removed (solids, fat and sand). The sewage passes through an automatic grid system, in the elevated station, in order to trap these thick objects so that there is no obstruction in the channel, and thus avoid damage to the system.

As the sewage gets to the well it goes through an automatic dam, which allows the effluent to get to the thick sewage bombing. The underground bombs press the sewage into the treatment plant. There is then the removal of organic matter. Eventually, after the addition of chlorine, the water becomes disinfected. The effluents then pass through the emissary, and on to their final destination, with a high level of purity.

Unitanks: The sewage treatment itself happens in the Unitank system, which is made up of 3 tanks in each row and it consists of a system of communicating vases. It can be divided into three steps – aeration, intermediary, and sedimentation. During the first phase, the effluent is sent to an aeration tank and is then sent to an ongoing aeration tank, finally it is sent to the third phase – sedimentation.

The process is utterly aerobic biological, and it is sludge activated. The sludge in the tanks are formed by aerobic bacteria, protozoa, fungus, among others. The activated sludge in the tanks are fed by the sewage that feeds the treatment. Through the metabolic reactions of the micro-organisms the removal of the biodegradable substances is completed, the result being the treated effluent.

Withdrawal of the sludge: In the final phase of the treatment, there is the appearance of the sludge, and the treated effluent is taken back to the river, through small channels and pipes. The resulting sludge is sent to the treatment process, where it is centrifuged.

The result of the whole process is that the water going back to the Itapemirim River is thoroughly cleaned, and safe.

In summary, the Fluxogram (phases) of the sewage treatment plant can be explained, as follows:

First, the residual water comes from homes, shops, etc.

Then, the residual water passes through small diameter pipes (secondary collectors). Then, the collector trunks – wider pipes, get the effluents from the secondary collectors. Then, the wide leading interceptor conductors take the effluents to the Sewage Treatment Station.

Next, the emissary takes the residual water from the station to the receptor body. At the Sewage Treatment Station there is a grid that removes the thick solids, which go to a sand box, for a secondary treatment, through a biological process.

As a result, there is a highly cleaned effluent.

The Foz do Brasil Projects

The Foz do Brasil conducts several environmental projects in order to Improve the indexes that performwater distribution and collection and sewage treatment in the city of Cachoeiro de Itapemirim.

The "Rio Vida Reflorescer" (Blooming River): One of these projects is the "Rio Vida Reflorescer" (Blooming River). The Project aims to recover the important areas for recharging the Itapemirim River water. About 15 thousand people have been reached through the cleaning of 15 water strings, which total 60 acres of recovered areas, with the plantation of 15 thousand of seedling units.

Waste Management and the Recycling of Reusable Material (NGO) Projects: The Foz do Brasil has implanted the concept of the

3Rs in the company (Reduction, Re-use and Recycling), involving 179 employees and 282 outsiders. The objectives of the projects are to educate those involved on environmental issues and the correct destination of the waste generated.

The results obtained from the projects have been a changing in the behavior of those involved and their family members, as well as the creation of new jobs and income.

United Group (Recycling of Reusable Materials NGO)

Environmental Educational Program – Rio Vida na Escola e Comunidade: Life in the River (Rio Vida) in Schools and Community – The program promotes changes in the behavior of society, aiming at the preservation of hydric resources.

The number of the directly affected public is over 47 thousand people, among them are students (pre-school, elementary school, high school and graduate students), teachers, communities, companies, class organizations and society in general, with the total number of individuals indirectly affected by the project being over 141 thousand.

Clean Streams: This is the company's main program, which aimed to reach a target of 90% of the volume of treated sewage by 2012. The program was initiated in 2006 and has treated 20km of streams that go through several sections in Cachoeiro de Itapemirim, and thus appreciating real estate buildings and improving the quality of life.

Good practices for the Sanitary System in Cachoeiro de Itapemirim: The program promotes the effective participation of the educational community in the process of sewage treatment of Cachoeiro de Itapemirim.

Getting to know my Company: It makes it possible to acquire knowledge on the working processes and programs developed by the company, as well as on the promotion of environmental education, aiming to achieve preservation and sustainability.

Social tariff: As an initiative of the local government, the company makes it possible for the population in need of access to essential water services and the sewage treatment for free, thereby benefiting 1036 families.

Advances in the Treatment

The number of households that receive sewage treatment has increased from 5% to 95.5% in the urban area. Sewage collection has gone from 75% to 97%, and the treated water supply has risen up to 100%, with 99.5% being in the municipality as a whole.

Water Treatment Station: Widening and updating of the capacity of the Water Treatment Station in the municipality, which has gone from 500 to 800 liters per second.

Advances in the water distribution system: Widening of the water distribution net to over 100 km.

Operational Improvement in the districts, which has made it possible to assure the population of those areas have a good quality supply of water.

Reduction of the water consumption average from 190.60 liters/ inhabitant per day, in 1998, to 121 liters, in 2009, as a result of educational actions developed in partnership with different sectors of the society.

Building of nine reservoirs, with a total capacity of 7.7 million liters. Therefore, as far as water distribution is concerned, Cachoeiro de Itapemirim has the same rank as the main Brazilian capitals.

Reduction in power consumption: The power that was used to operate the company was reduced from 1,035,749 kWh, in 1999, to 852,266 kWh, in 2009.

Modern technology and infrastructure: The on-line telemetry, in the company's main office, allows the monitoring, telecontrol, and telecommand of the main functions in the system from over 30 operational units.

Laboratories with precise and state-of-the-art equipment. Hydrometering Index of 99%.

The building of a Sewage Treatment Station with a capacity of 560 liters per second.

Advances in the collection and sewage treatment: The operation of 12 km interceptors, 10 km of trunk collectors, and 88 km of sewage collection.

The withdrawal of 21 million liters of sewage which, up to that time, were directly dumped into the Itapemirim River.

Collection of 97% of all the sewage in the city and treatment of 87%. The company has invested R$ 60 million in Cachoeiro de Itapemirim.

In ten years, the company has invested R$ 32.00 per capita, which represents a higher rate than the Brazilian average, which is R$ 19.00 per capita.

After eleven years' concession, the company has presented the following comparative number before the concession, in 1998: 75% sewage collection in town; Manual measuring system; three days' average time to attend the client; Water Treatment Station with a capacity to process 500 liters' water per second; treatment of 5% of the collected sewage; 95% of attendance with the water net in the urban area; Water average consumption in the county of 190,60 liters per inhabitant; Hydrometer index of 76% – capacity to correctly measure household consumption.

In 2009, the progress was as follows:

Collection of 97% of all the sewage in the town – Computerized measure system.

Average time to attend the client: 12 hours

Water Treatment Station with the capacity to process 800 liters water per second

87% treatment of the collected sewage

Water attendance in the urban area of 99.99%.

Average water consumption in the municipality of 121 liters per person

Hydrometer index of 99%

Poll

Polls inform the satisfaction rate of the client in relation to the services performed in town. The most recent, conducted by the Futura Institute, shows a 79% satisfaction rate.

The company will invest R$ 75 million in Cachoeiro de Itapemirim.

In five years time, the company will invest R$75.00 per inhabitant, a higher index than the Brazilian average, which will be R$ 22.00.

The New Investment Programs include the implantation of the PHC – a small hydroelectric plant of 2.8 MW, which is a project to auto-generate electric Power in order to be highly autonomous with the supply of water services.

Future Challenges

- Continuously investing on the widening and the operational updating of the water supply systems and of the collection and treatment of the sewage;
- Widening the collection and implementation of sewage treatment in the districts;
- Optimizing the buildings of the Clean Water Stream Program, the target being to finish by 2011;
- Reaching the ISSO 14001, and the ISO IEC 17025;
- Widening the Environment Educational Program;
- Starting the Small Hydroelectric Central (PCH), which will make the company self- sufficient in terms of power.

References

INSTITUTO BRASILEIRO DE GEOGRAFIA E ESTATÍSTICA. *PNSB 2008*: abastecimento de água chega a 99,4% dos municípios, coleto de lixo a 100%, e rede de esgoto a 55,2%. Available at: *http:// www.ibge.gov.br/home/presidencia/noiticias/noitica_visualiza.php?id_noticia=1691&id_pagina=1*. Access on: June 25, 2010.

MILLENNIUM Project. *2002-2006*: commisioned by the UN Secretary General and Supported by the UN Development Group. Available at: http://www.unmillenniumproject.org/goals/gti.htm. Access on: June 29, 2010.

PERCEPÇÕES sobre saneamento básico. *Trata Brasil*. Available at: *http://www.tratabrasil.org.br/novo_sitems/APRESENTACAO_*

IBOPE_TRATA/Apresentacao_coletiva_12ago.pdf.. Access on: June 25, 2010.

RANKING do saneamento. *Trata Brasil avalia os serviços prestados nas 81 maiores cidades do Brasil*. Trata Brasil. Available at: *http://www.tratabrasil.org.br/novo_site/?id=12089*. Access on: June 25, 2010.

THIELMANN, Beatriz. *Esgoto doméstico é a maior fonte de poluição do Rio Tietê*. Globo Repórter. July 17, 2009. Navega São Paulo: educação e vida pelo Tietê. Available at: *http://www.navegasp.com.br/index.jsp?conteudo=161*. Access on: June 29, 2010.

20

Using Role Play to Build a Common Sustainable Development Agenda for Brazil-China: the Case of Top China Program

Rogério César Pereira de Araújo[1]
Leda Nakashiro Makiya[2]

Introduction

According to the Brandtland report, sustainable development aims to promote the improved well-being of the current generation without undermining the capacity of future generations to reach at least the same well-being level as previous generations (UN, 1987). As a fundamental principle, the sustainable development strategies must balance the economic, social and environmental goals. This will not be achieved without all stakeholders engaging in the public decision-making process (Buchs; Blanchard, 2011).

Sustainable development is a challenge for China and Brazil alike. Recently, each country has undertaken considerable changes in the economy aiming to promote economic growth and being able to increase living standards for their respective populations. However, rapid economic growth does not happen without negative impacts on the environment. Such a context raises concerns on whether the

1 Universidade Federal do Ceará, Fortaleza, Ceará, Brazil. E-mail: *rcpa@ufc.br*.

2 Universidade Estadual de Campinas, Campinas, São Paulo, Brazil. E-mail: *iedakm@gmail.com*.

path taken by both countries can lead them to a point of no return regarding the loss of environmental services. In order to avoid the undesirable consequences, not only knowledge and technology are necessary, but also building political capital that takes the ecosystem and its immutable laws into consideration. Building political capital requires educating the current and future decision makers on the fundamental concept related to nature as well as social interaction, integration, interdependence and cooperation.

Traditional teaching techniques (e.g., lecture) seem insufficient to account for the complexity of the global and environmental problems and the ethical principles embodied within sustainable development principles (Doorn; Kroesen, 2011). For this reason, the role play was used approach so that understanding of the sustainable development concept could be developed. The group dynamics consisted of making the students from China and Brazil assume the role of relevant stakeholders in each country and allow them to think and act according to their interests. Based on these interests, the government officials of both countries engaged in negotiation in order to reach agreements and form a common agenda to address policies directed towards the sustainable development of both countries.

Specifically, the group dynamics aimed to achieve the following goals:

- Identify, discuss and propose solutions to relevant problems concerning sustainable development in China and Brazil;
- Promote interaction among students from China and Brazil;
- Understand the political and negotiation processes involving stakeholders interested in economic growth and sustainable development;
- Identify complementarities that could motive cooperative agreements between China and Brazil.

This group dynamics were conducted by two Brazilian professors participating in the 2011 Top China in July 7th, 2011, at Peking University.

Group Dynamics

This section presents the assumptions and procedures used in the group dynamics. The assumptions are concerned with the current political system in Brazil and China. The procedures consist of presenting and defining the types of stakeholders involved, forming groups among students, assigning the groups' roles, setting up the activity schedule and time limits, and showing the guideline principles for the success of the group dynamics.

Political System and Stakeholders Definition

The group dynamics assume the current political setting in Brazil and China. Based on the political structure of each country, relevant stakeholders are selected and assigned to the groups.

Brazil is a democratic country and the governmental sphere is divided into three political powers: executive, legislative, and judiciary. The citizens organize themselves in political parties that dispute political power through an elective process to run the government at municipal, state and federal levels. The nongovernmental organizations and diverse groups of interest also have important political roles in Brazil. For the group dynamics, five stakeholder types was selected, which were as follows: federal government officials, private entrepreneurs, environmentalists, vulnerable group activists, and indigenous people's leaders. The Brazilian stakeholders' responsibilities and interests are presented in Table 1.

Table 1. Brazilian political system defined for the group dynamics.

Stakeholder	Description
1. Brazilian Government Officials	The government is responsible for proposing and carrying out economic, social and environmental policies in order to improve the population's living standards and promote economic growth and development. The government is also responsible for identifying opportunities to establish economic, social and cultural links with other countries.
2. Private Entrepreneurs	The objective of firms is to obtain profit through the producing and selling if goods in the domestic and external markets.

Stakeholder	Description
3. Environmentalists	The environmentalists take actions to protect the environment, biodiversity and relevant ecosystems as well as the sustainable use of the country's natural resources. They believe in the integrated management of the ecological and economic systems in order to achieve sustainable development.
4. Vulnerable Group Activists	The vulnerable population is constituted by poor people who depend directly on natural resource and face difficulties to meet their basic needs in terms of housing, food, health and education, etc. They include peasants in rural areas, traditional communities in the (rain) forest and coastal zones.
5. Indigenous People's Leaders	The leaders are responsible for representing indigenous peoples from Brazil and defending their rights for land, livelihood and environment.

China is a communist country which is governed by only one political party. The political structure in China is complex and, for the group dynamics proposes, five stakeholders were considered: central government officials, local government officials, multinational entrepreneurs, environmental agency officials, and community and village leaders. This political structure was defined with the help of Chinese students. The Chinese stakeholders' characteristics and interests are presented in Table 2.

Table 2. Chinese political system defined for the group dynamics.

Stakeholder	Description
1. Central Government Officials	Central government is responsible for planning and carrying out policies to develop the economic, social and environmental sectors in order to improve the people's living standards. The government is also responsible for identifying opportunities to establish economic, social and cultural links with other countries.

Stakeholder	Description
2. Local Government Officials	Local government is responsible for executing the central governmental policies and is also empowered to take its own economic decisions to develop the local economy.
3. Multinational Entrepreneurs	Multinational firms are interested in investing in China's economy and supplying goods to meet the domestic and external market demands while benefiting from the Chinese comparative advantage in the global economy.
4. Environmental Agencies' Officials	The environmental agencies take actions to protect the environment, biodiversity and relevant ecosystems and the sustainable use of the country's natural resources. They believe in the integrated management of the ecological and economic systems in order to achieve sustainable development.
5. Village/ Community's Leaders	The village/community leaders are responsible for identifying the people's wants and needs and communicate them to the Local Government Officials.

Four Brazilian professors were also part of the group dynamics who took the role of international multilateral observers. Their roles were to observe, mediate and evaluate the interaction among the countries' stakeholders. The description of the international observers' role is provided in Table 3.

Table 3. International observers.

Stakeholder	Description
International Multilateral Observers	International observers are interested in mediating and supporting agreements among countries in order to promote international cooperation, integration and global sustainable development. The officials are responsible for evaluating and endorsing the cooperation agreements among the countries.

Procedures: steps and time limit

The group dynamics began with the instructors presenting the fundamental concept and the tasks to be accomplished by the participants. The meaning of sustainable development, globalization, and the environmental problems faced by the countries around the world were discussed. After this, each step that students had to follow in order to achieve the goals of the group dynamics were outlined and explained. Table 4 shows these steps, their description and the time limit for each activity.

Table 4. Steps of the group dynamics.

Activity	Description	Time
Introduction	Presenting the context and definition of sustainable development, globalization and environmental crisis in the World.	10 min
Group dynamics objective and procedures	Presenting the group dynamics' structure, steps and time limits.	10 min.
Groups' formation	Distributing the students into groups and assigning roles to them.	10 min.
Group work and reports	Working together in the group to identify at least three priorities (problems, solutions and opportunities) to meet the group's interest and future prospects.	20 min.
Group presentation	Presenting the outcomes of the group work by the leader or group members.	10 min. each
Governments' meeting	Holding a meeting between the Chinese and Brazilian senior government officials to define a common agenda of cooperation to meet the stakeholders' interests and the sustainable development goals.	20 min.
International observers' evaluation	Evaluating the output presented by the stakeholder groups and establishing future prospects for the sustainable development through cooperation between China and Brazil.	20 min.

Activity	Description	Time
Plenary	Presenting and discussing the common agenda for sustainable development in China and Brazil.	30 min.
	Total	3 hours

Before engaging in the group dynamics itself, the students were distributed into groups: The Brazilian students played the role of Brazilian stakeholders and Chinese students played the role of Chinese stakeholders. In principle, each group assumed the commitment of pursuing the interests and goals of the stakeholders in the hypothetical political setting.

Figure 1 presents the scheme of activity sequence conducted in the group dynamics. Some activities occurred simultaneously, such as the work performed by the stakeholder groups; the senior government official meeting and the international observers' evaluation. During the stakeholder group presentations stage, the senior government official group listened and took notes on the stakeholders' demands. These demands would then be discussed at the senior government officials' meeting. Towards the end of the section, the government official groups made a final pronouncement at the Plenary answering the demands brought up by the stakeholders' group.

Figure 1. Group dynamics activities.

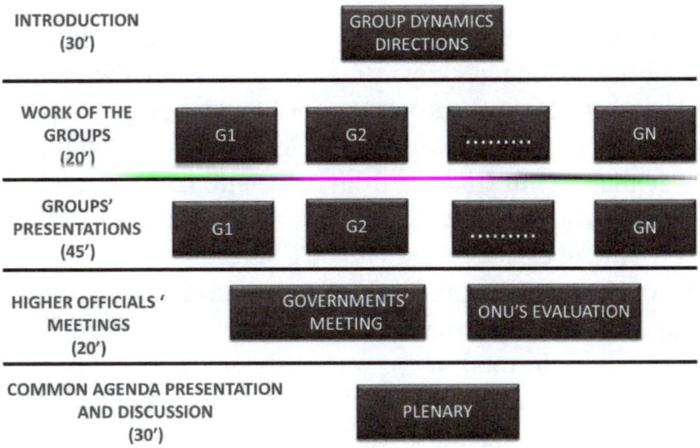

Guiding principles

The success of the group dynamics depends on various factors such as the level of information held by the participants and how the groups organize themselves to accomplish the task. In order to facilitate the groups carrying out the tasks, it was suggested that each group elect a leader and annotator. The former would be responsible for assuring that the group works as a team and guaranteeing that all members of the group have a voice. The annotator would be responsible for controlling the time and guaranteeing that the tasks are completed on time. The importance of participation and interaction among the group members for the success of the group dynamics was also emphasized. In addition, it was stressed that the time limit should be rigorously obeyed and the group presentation be clear, objective and concise.

Finally, a set of guiding principles to govern the groups' activities and their interaction was set up: (i) all members must participate in the discussion; (ii) differing opinions must be respected; (iii) mutual cooperation must be encouraged among members; (iv) each member should speak one at a time; (v) members must not speak in a loud voice; and (vi) all groups must respect the time limit.

Results

Group Dynamics: Numbers and Pictures

The group dynamics was conducted with five Chinese groups involving fifteen students, five Brazilian groups involving thirty two students, and one group made ups by four Brazilian professors assuming the role of international observers. The group was assembled according the availability of students from both nationalities. The Brazilian stakeholder groups had a relatively larger number of students per group (7-8 members) compared to the Chinese stakeholder groups (3 members). However, the difference in group size did not significantly affect the groups' performances, as can verify by the quality of the groups' outputs. In total, there were fifty-seven participants in the group dynamics. Table 5 shows the stakeholder groups by country and the number of students in each group.

Table 5. Number of groups and students participating in the group dynamics.

Country	Group	Stakeholder	Number of students
China	1	Central Government Officials	3
	2	Local Government Officials	3
	3	Multinational Entrepreneur	3
	4	Environmental Agencies' Officials	3
	5	Village/Community's Leaders	3
Brazil	1	Brazil Government Officials	8
	2	Private Entrepreneurs (Firms)	8
	3	Environmentalists (NGOs)	7
	4	Vulnerable Groups Activists	7
	5	Indigenous People Leaders	8
International	1	International Multilateral Observers	4
Total			57

Figures 2 to 7 show pictures taken when the students were participating in the group dynamics. Figure 2 and 3 show a Brazilian stakeholder group and a Chinese stakeholder group, respectively, working on identifying problems, solutions or opportunities concerning their stakeholder's interests and goals.

Figure 2. Brazilian stakeholder group.

Figure 3. Chinese stakeholder group.

Figure 4. Chinese group presentation.

Figure 5. Brazilian group presentation.

Figure 6. Senior government officials' negotiation meeting.

Figure 7. International observers' evaluation.

Figure 4 and 5 show the Brazilian and Chinese stakeholders presenting the results of their group work. Figure 6 shows the senior governmental officials in the negotiation meeting building a common sustainable development agenda for Brazil-China. Finally, Figure 7 shows the international observers evaluating the results presented by the stakeholder groups.

Results of the Group Dynamics

The results of the Chinese and Brazilian stakeholder groups are presented and compared at each level, as shown in Table 6, beginning with the most vulnerable stakeholder declarations and finishing with the senior government officials' statements.

Table 6. Number of groups and students participating in the group dynamics.

China	Brazil
Village/Community's Leaders	Indigenous People's Leaders Vulnerable Group Activists

China	Brazil
Environmental Agencies' Officials	Environmentalists
Multinational Entrepreneurs	Private Entrepreneurs
Local Government Officials Central Government Officials	Brazilian Government Officials

Vulnerable Stakeholders

- Chinese Village/Community's Leaders:

This group emphasized the economic, social and environmental aspects affecting the stakeholders. For the economy, it was stressed that "the people are not as rich as those who live in big cities in China. The proposed solution for this problem is to attract more investment to build more factories to provide more job opportunities". For society, the group pointed out that "people do not receive as much education as their counterparts in big cities. The proposed solution for this problem would be the allocation of more investment by the government to improve education in rural areas". Concerning the environment, it was explained that "the rubbish from big cities is buried in the suburbs or near villages. The villagers do not have trash classification. understanding The proposed solution for this issue is to force the government to make regulations to deal with the rubbish from big cities. It should also to put more emphasis on imposing the awareness of villagers regarding trash classification".

- Brazilian Vulnerable Group Activists:

This stakeholder group pointed out that the vulnerable group activists firstly demanded "being heard". They raised the housing problems as one that needs a quick solution, mainly concerning the increase of its availability and its proximity to the work places. The group also discussed the need for better urban sanitary conditions and health care. At the same level of importance, they talked about increasing access to green parks and providing environmental education for the population.

- Brazilian Indigenous People's Leaders:

This stakeholder group made demands about land reform, the environmental impact caused by infrastructure development, governmental intervention and quality of the health and education services. Concerning land reform, the stakeholder group expressed in their own words: “we want our land to be demarcated, so we can preserve our tradition”. They affirmed that “it is important to recognize our connection with the place in which we live. It is our right to stay in the places where we and our ancestors were born”. They also talked about the huge problem they are facing as a result of the construction of the Belo Monte Hydroelectric Power Station. They stressed the eminent risk faced by the communities near Xingu River, such their relocation to new areas. They said “We do not want this!”. They mentioned the damage caused to the environment, such as the “extinction of species (animals and plants), raising the number of diseases due to changing the river course, and increasing the level of starvation among indigenous people since we feed our family by extracting products from the ecosystem”. They protested loudly about stopping the construction saying “We ask that BELO MONTE BE STOPPED NOW! We strongly demand our rights to be respected”.

Environmental Stakeholders

- Chinese Environmental Agency Officials:

The environmental agency officials specifically demanded augmenting the financial and human resource in order to allow them to accomplish their mission of enforcing the environmental law and improving environmental performance for the whole country.

- Brazilian Environmentalists:

The stakeholder group raised awareness about deforestation, energy and the lack of environmental education in Brazil. Concerning deforestation, they demanded stricter enforcement by increasing the number of inspections, systematically rotating the local inspectors, providing better salaries and working conditions to curb corruption and setting up satellite monitoring (such as CBERS) integrated to the land enforcement framework. Regarding the energy sector, they demanded an increase in alternative energy generation and improvement in efficient energy use, such as promoting the production and consumption of energy efficient electronic goods and providing incentives to companies that join this program. They also pointed out the need for public campaigns in order to disseminate information about saving energy and water, recycling and sustainable waste management in order to raise public awareness concerning environmental issues in Brazil.

Local Government and Entrepreneurs Stakeholders

- Chinese Local Government Officials:

This stakeholder group asked for a higher responsibility from the Central government, an increase in job opportunities at local communities and leveraging of investment in high technology industries. in addition, they also demanded stricter rules for environmental protection.

- Chinese Multinational Entrepreneurs:

The Chinese multinational entrepreneurs made only one request that was reduced taxation on the multinational firms upon the understanding that they assume an important role for Chinese technological and trade development.

- Brazilian Private Entrepreneurs:

The private entrepreneur group asked for better conditions to build a bilateral agreement in technological areas with the aim of improving products and market values in Brazil and China. They also pointed out the need to reduce taxes on Chinese products.

Senior Government Official Stakeholders

- Chinese Central Government Officials:

The Central government officials group, taking into consideration all the stakeholders' opinions, decided to take the following actions (using the words of the group members): (i) "Now we are aware of the environmental agency's enthusiasm and willingness to supervise the environment protection work. We did not pay enough attention to them, now we decide to establish a regular mechanism that those agencies can submit their comments on the main factories a quarter a time. To those who submit a high quality and realistic report, we can give financial support and reward according to the quality level of the report; (ii) The representative of the factories showed a strong willingness to reduce the taxation burden on them. After consideration, we have decided to make a new taxation level system. In this system, the factories are divided into different levels, which depends on the environmental investment they have made. For those who have their own sewage treatment plants, air purifier or those that can help solve the pollution emission, we will apply reduced taxation; (iii) Considering that China is a major developing country with a vast territory, huge population, weak infrastructure, there are significant differences in different regions that need to be addressed. So, the central government policy will only act as a guide, not as specific instructors of a certain area of policy. As long as the direction is correct, central government will try its best to give enough space so that local governments can make their own development plans; (v) The other important part is to negotiate with the Brazil government. The two sides have a common opinion about each other's profit; (iv) As China is developing so quickly, China needs strong support from Brazil regarding natural capital and energy resources. At the same time, China should provide more support for labor resources, technology of infrastructure. Only if the both sides benefit from the corporation can we continue developing and improving the international relationship between the countries; and (vi) We do not understand each other sufficiently in terms of education and technology. It is necessary for both sides to establish a mechanism to make regular, frequent communication possible. As for these topics we have both addressed, such as new energy, the problems in the urbanization, the balance of rapid development and pollution, we need to share what we have learnt from each other's experience. Involving the young generation within the approach to find solutions will be a long-term task for each government so that there can be sustainable control of these problems".

- Brazilian Government Officials:

The Brazilian government officials group expressed its intentions to create bilateral standards for products at final consumer and industrial levels. They also emphasized the need to establish commercial agreements concerning the exchange rate of Brazilian and Chinese currencies (RMB and Real). The Brazilian government officials pointed out the need to increase investment and technology exchange, mainly in realation to public transport (train, electric transportation) and renewable energy (bioenergy, hydroelectric power). Another area that needs cooperation is the implementation of agricultural technology for dry areas in order to forge rural development in those areas. The need to improve the higher education relationship between Brazil and China was also stressed, for example augmenting access to international scholarships for students and professors between Brazil and China.

Final Considerations

This group dynamics made it possible to capture the Brazilian and Chinese perception concerning sustainable development issues in their home countries, as well as how interaction and integration among multinational stakeholders can help to build cooperation among countries. In addition, it allowed students to play the role of relevant stakeholders, understanding their motivational behavior, and perceiving them as a political actor in an international negotiation setting. Finally, they had the opportunity to verify that confronting differing and conflicting views from around the world is a way to provoke critical thinking and promote the comprehensive education that is necessary to build sustainable relationship among nations.

References

BUCHS, A.; BLANCHARD, O. Exploring the Concept of Sustainable Development Through a Role-Play. *Journal of Economic Education*, v.42, n.4, p.388 97, 2011. DOI: 10.1080/00220485.2011.606089.

DOORN, N.; KROESEN, J. O. Using and Developing Role Play in Teaching Aimed at Preparing for Social Responsibility. *Sci. Eng. Ethics*, 20 of December 2011. DOI 10.1007/s11948-011-9335-6.

UN – UNITED NATIONS. *Our Common Future*: Report of the World Commission on Environment and Development. United Nation, 1987.

21

Three-Dimensional Air Quality Models

Ping Liu

History and Current Status of Three-Dimensional Air Quality Models

Ambient measurements and numerical models are the approaches that are widely used to understand the transformation and transport of air pollutants in the atmosphere, however, all of these have their own limitations. The measurements are usually limited due to fixed times and locations, while the Three-Dimensional (3-D) air quality models can continuously simulate numerous physical and chemical processes that will affect the evolutions of the air pollutants in the "whole" atmospheric system. Such simulations can help us understand the interactions of different processes, to design measurements, and to make regulatory decisions. The evolutions of regional 3-D air quality models used in the U.S. are introduced and reviewed here to provide the general background for the models used in this study.

The 3-D air quality models were initially applied to simulate gas-phase pollutants such as O_3 and sulfur dioxide (SO_2) in the early 1970s to 1980s. The first 3-D model, the Urban Airshed Model (UAM), was developed by John Seinfeld's group at Caltech for photochemical air pollution and applied to the Los Angeles airshed (Reynolds; Roth; Seinfeld, 1973). The treated processes included emissions, transport, and chemical transformation. These processes only consisted of 15 reactions involving 10 chemical species, and focused on carbon monoxide (CO), nitric oxide (NO), nitrogen dioxide (NO_2), hydrocarbon and O_3 (Reynolds; Roth; Seinfeld, 1973). The Sulfur Transport Eulerian Model version 1 (STEM-I) was developed to simulate the regional distribution of SO_2 and sulfate in the lower troposphere (Carmichael; Peters, 1984a), and was applied to the eastern United States to study the regional transport of sulphur oxides (SO_x) (Carmichael; Peters, 1984b). An SO_2 photochemical oxidation mechanism with 27 reactions was used, which also considered the SO_2 oxidation in an aqueous-phase, dry deposition, and wet deposition (Carmichael; Peters, 1984a). The Regional Oxidant Model (ROM) was used to simulate regional O_3 and other species (Lamb, 1982; Schere; Possiel, 1984). All the above air quality models use analyzed or calculated meteorological inputs from observations, and treat transport, emissions, deposition, and simple chemistry. These models have continuously evolved, taking into account scientific advances.

From the 1980s to the early 1990s, 3-D air quality models were applied to simulate O_3 and acid deposition, which can use analyzed or simulated meteorological inputs, treat cloud and precipitation scavenging, and include detailed chemistry. Carbon Bond Mechanism version 4 (CBM-IV) was incorporated into the UAM for solving chemical kinetics (Gery; Whitten; Killus, 1988). The wind inputs were defined either by interpolating surface and upper-air observations, or through the use of a diagnostic wind model with interpolative techniques in UAM (Scheffe; Morris, 1993). The Regional Acid Deposition Model (RADM) incorporated the RADM2 gas-phase mechanism (Stockwell, 1986), aqueous-phase reactions, and also considered cloud processes. This Model was applied to calculate episodic chemical concentrations and dry/wet depositions of acids in North America (Chang et al., 1987). The Sulfur Transport Eulerian Model version 2 (STEM-II) was improved to include detailed gas-phase (Atkinson; Lloyd; Winges, 1982) and aqueous-phase chemistry, the scavenging by clouds and precipitation, and was applied to various urban, regional, and mesoscale acid deposition studies (Carmichael; Peters; Saylor, 1991). The photochemical oxidant model, CALGRID, was updated from the UAM and developed for regulatory assessments of O_3 control strategies (Yamartino et al., 1992).

From the middle 1990s to the present day, 3-D air quality models have been widely used to simulate O_3, PM, regional haze, and air toxics. The models use meteorological fields simulated by meteorological models such as the Pennsylvania State University (PSU) / National Center for Atmospheric Research (NCAR) Mesoscale Modeling System Generation 5 (MM5) (Grell; Dudia; Stauffer, 1995). The models also include size/chemical resolved PM, detailed aerosol microphysics, and detailed mixed-phase chemistry. The Urban and Regional Multiscale Model (URM) (Kumar, 1994) was applied to Southern California and included fine scales within the coarse grid resolution to capture the details of the atmospheric processes (Kumar; Odman; Russell, 1994). The Regional Particulate Matter Model (RPM) was modified based on RADM and added particle treatments to simulate aerosol concentrations and particle size distributions over eastern North America (Binkowski; Shankar, 1995). The Gas, Aerosol, Transport, and Radiation (GATOR) Eulerian air quality model has detailed aerosol treatments and was applied during the Southern California Air Quality Study (Jacobson et al., 1996; Jacobson, 1997a, 1997b). Particularly, GATOR is coupled interactively to a prognostic meteorological model, the Mesoscale Meteorological and Tracer Dispersion Model (MMTD), and GATOR feeds the radiation calculation back to the MMTD model (Jacobson et al., 1996; Jacobson, 1997a, 1997b). The California/Carnegie-Mellon Institute of Technology (CIT) air quality model is a size-resolved and chemically resolved aerosol model, and was applied to simulate gases and aerosols in the south coast Air Basin of California (Meng; Dabdub; Seinfeld, 1998). In CIT, the gas-to-particle conversions were represented by dynamic mass transfer between the gas and aerosol phases. The U.S. Environmental Protection Agency (EPA)'s third generation air quality model, the Community Multiscale Air Quality Modeling System (CMAQ), represents the concept of "one-atmosphere" (Byun; Ching, 1999), treats multiple pollutants (i.e., O_3, PM, visibility, and air toxics) and can be used for multiple scales such as regional and urban. It considers the major processes that occur in the atmosphere: transport, emissions, gas-phase chemistry, aqueous-phase chemistry, dry and wet depositions, and aerosol dynamics (Byun; Ching, 1999; Binkowski; Roselle, 2003; Byun; Schere, 2007). The particle size distribution uses three lognormal subdistributions, and the processes of nucleation, coagulation, and condensation are all considered. Several detailed gas-phase mechanisms are available including CBM-IV, the updated and expanded version of the carbon bond mechanism (CB05)

(Sarwar et al., 2008) and the Statewide Air Pollution Research Center Mechanism (SAPRC99) (Carter, 2000) are used to represent complex and detailed atmospheric chemistry. The Comprehensive Air Quality Model with extensions (CAMx) (ENVIRON, 2004) and the Multiscale Air Quality Simulation Platform (MAQSIP) (Mathur et al., 2005) feature a "one-atmosphere" treatment for O_3 and PM as well, with detailed algorithms for the relevant scientific processes. In particular, the CMAx has a two-way nested grid structure that allows it to be simultaneously run with coarse and fine grid resolutions.

From the early 2000s to the present day, air quality models have been improved by coupling online-chemistry with a meteorological model. The first online meteorology/chemistry/aerosol/radiation model is the described previously GATOR/MMTD (Jacobson et al., 1996; Jacobson, 1997a, 1997b). Another integrated model for atmospheric meteorology, physics, and chemistry with online chemistry coupled into a new state-of-the-art meteorological model, the Weather Research and Forecasting (WRF) model (http://www.mmm.ucar.edu/wrf/users/document.html), was developed and named WRF/Chem (Grell et al., 2005). WRF/Chem uses dynamically estimated biogenic emissions with meteorological inputs from the WRF model, online calculated photolysis frequencies, and considers aerosol feedback to the meteorological simulation. It also provides a two-way nesting function. WRF was also integrated with CMAQ, which was a two-way coupled system that can simulate aerosol feedback (Wong et al., 2012).

The 3-D air quality models have been applied to various regions and episodic studies in the U.S. Several evaluations have been conducted for most models such as the U.S. EPA's CMAQ model (Eder; Yu, 2006; Appel et al., 2007). Comparisons among the performances of different models have also been conducted (e.g., Biswas et al., 2001; Hogrefe et al., 2001; Zhang; Pun; Wu, 2004; Morris et al., 2006; Hogrefe et al., 2006; Bailey et al., 2007; Pun et al., 2008). There are several comprehensive reviews of regional and urban scale air quality models in terms of their formulation, similarities, and differences (Peters et al., 1995; Russell, 1997; Russell; Dennis, 2000; Seigneur, 2001), including a new review of online coupled meteorology and chemistry models (Zhang, 2008). Generally, the 3-D air quality models perform well for O_3, and fine particles ($PM_{2.5}$) mass concentrations, but have limitations regarding their treatment of aerosols (Seigneur et al., 1999; Russell;

Dennis, 2000; Seigneur, 2001). The areas for improvement include more reliable emissions and meteorological fields, cloud processes, and the treatment of secondary PM (e.g., sulfate, nitrate, and secondary organic aerosols) formation (Seigneur et al., 1999; Seigneur, 2001; Wayland et al., 2002). Lastly, no one model is able to provide the best treatments of all processes; balance between increased model details and computational intensity needs to be archived (Russell, 1997).

Applications of 3-D air quality models in Shanghai

The CMAQ model (v4.6) was applied from 19 January to 17 February 2009 in the Shanghai episode, with a 3-km horizontal grid resolution. The simulated concentrations of SO_2, NO_2, O_3, coarse particles (PM_{10}), and $PM_{2.5}$ were compared with the observations from the routine monitoring network and special monitoring sites in Shanghai during this period. The comparisons showed that the CMAQ model overpredicted O_3, SO_2, NO_2, and $PM_{2.5}$ by 13.1%, 19.6%, 35.3%, and 1.4%, respectively, and underpredicted PM_{10} by 1.7%. Despite the CMAQ model being able to simulate PM_{10}, $PM_{2.5}$, SO_2, NO_2, and O_3 with relatively low biases, it had limitations in predicting hourly variations of these pollutants. The diagnostic tool embedded in the CMAQ, process analysis (PA), was conducted to identify the contributions of major atmospheric processes regarding the formation and removal of the major air pollutants. The PA results showed that emissions were the major contributor to the increases of PM_{10}, SO_2, and NO_2 at the urban site, and transport was the major contributor to the changes of those pollutants at the rural site. The model biases were probably due to the uncertainties in the meteorological simulations (e.g., terrain, cold bias of temperature) and emissions (e.g., emission inventory, uncounted emission such as fireworks). This requires that there be further testing and improvement of the model inputs and the model treatment.

Acknowledgements

The applications of the CMAQ model in Shanghai was supported by National Natural Science Foundation of China (41005077) and

Doctoral Fund of Ministry of Education of China (20100073120042). With thanks to Qian Wang and Qingyan Fu from Shanghai Environmental Monitoring Center for providing the observational data.

References

APPEL, W. K. et al. Evaluation of the Community Multiscale Air Quality (CMAQ) model version 4.5: sensitivities impacting model performance part I-ozone. *Atmos. Environ.*, 41, p.9603-15, 2007.

ATKINSON, R.; LLOYD, A. C.; WINGES, L. An updated chemical mechanism for hydrocarbon/NO_x/SO_2 photooxidations suitable for inclusion in atmospheric simulation models. *Atmos. Environ.*, 16, p.1341-55, 1982.

BAILEY, E. M. et al. A comparison of the performance of four air quality models for the Southern Oxidants Study episode in July 1999. *J. Geophys. Res.*, 112, 2007. D05306, doi: 10.1029/2005JD007021.

BINKOWSKI, F.; SHANKAR, U. The regional particulate matter model 1. model description and preliminary results. *J. Geophys. Res.*, 100(D12), p.26191-209, 1995.

______.; ROSELLE, S. J. Models-3 community multiscale air quality (CMAQ) model aerosol component, 1. model description. *J. Geophys. Res.*, 108, p.4183, 2003. doi:10.1029/2001JD001409.

BISWAS, J. et al. Evaluating the performance of regional-scale photochemical modeling systems; part III-precursor predictions. *Atmos. Environ.*, 35, p.6129-49, 2001.

BYUN, D. W.; CHING, J. K. S. Science algorithms of the EPA Models-3 community multiscale air quality (CMAQ) modeling system, EPA/600/R-99/030, Office of Research and Development, U.S. Environmental Protection Agency. Washington, D.C., 1999.

BYUN, D. W.; SCHERE, K. L. Review of the governing equations, computational algorithms, and other components of the Models-3 Community Multiscale Air Quality (CMAQ) modeling system. *Appl. Mech. Rev.*, 59, p.51-77, 2007.

CARMICHAEL, G. R.; PETERS, L. K. An Eulerian transport/transformation/removal model for SO_2 and sulfate I. model development. *Atmos. Environ.*, 18(5), p.937-52, 1984a.

CARMICHAEL, G. R.; PETERS, L. K. An Eulerian transport/transformation/removal model for SO_2 and sulfate II. model calculation of SO_x transport in the Eastern United States. *Atmos. Environ.*, 18(5), p.953-67, 1984b.

CARMICHAEL, G. R.; PETERS, L. K.; SAYLOR, R. D. The STEM-II regional scale acid deposition and photochemical oxidant model I. an overview of model development and applications. *Atmos. Environ.*, 25A(10), p.2077-90, 1991.

CARTER, W. P. L. Implementation of the SAPRC-99 chemical mechanism into the Models-3 Framework, report to the US Environmental Agency, 2000.

CHANG, J. S. et al. A three-dimensional Eulerian acid deposition model: physical concepts and formulation. *J. Geophys. Res.*, 92(D12), p.14681-700, 1987.

EDER, B.; YU, S. A performance evaluation of the 2004 release of Models-3 CMAQ, *Atmos. Environ.*, 40, p.4811-24, 2006.

ENVIRON. Comprehensive air quality model with extensions (CAMx), version 4.10s. User's guide, ENVIRON International Corporation, Novato, CA, 120 p., 2004.

GERY M. W.; WHITTEN, G. Z.; KILLUS, J. P. Development and testing of the CBM-IV for urban and regional modeling, [SYSAPP-88/002], Systems Applications International, San Rafael, CA, 1988.

GRELL, G. A.; DUDIA, J.; STAUFFER, D. R. A description of the fifth-generation Penn State/NCAR Mesoscale Model (MM5), NCAR Technical Note, NCAR/TN-398+STR, National Center for Atmospheric Research, Boulder, CO, 1995.

GRELL, G. A. et al. Fully coupled "online" chemistry within the WRF model. *Atmos. Environ.*, 39, p.6957-75, 2005.

HOGREFE, C. et al. Temporal features in observed and simulated meteorology and air quality over the Eastern United States. *Atmos. Environ.*, 40, p.5041-55, 2006.

HOGREFE, C. et al. Evaluating the performance of regional-scale photochemical modeling systems: part II-ozone predictions. *Atmos. Environ.*, 35, p.4175-88, 2001.

JACOBSON, M. Z. Development and application of a new air pollution modeling system-part II. aerosol module structure and design. *Atmos. Environ.*, 31(2), p.131-44, 1997a.

JACOBSON, M. Z. Development and application of a new air pollution modeling system-part III. aerosol-phase simulations. *Atmos. Environ.*, 31(4), p.587-608, 1997b.

JACOBSON, M. Z. et al. Development and application of a new air pollution modeling system-part I. Gas-phase simulations. *Atmos. Environ.*, 30(12), p.1939-63, 1996.

KUMAR, N. Multiscale modeling of urban and regional photochemical air pollution, Ph.D. dissertation. Carnegie Mellon University, 1994.

KUMAR, N.; ODMAN, M. T.; RUSSELL, A. G. Multiscale air quality modeling: application to Southern California. *J. Geophys. Res.*, 99(D3), p.5385-97, 1994.

LAMB, R. G. A regional-scale (1000 km) model of photochemical air pollution: part I. theoretical formulation, U. S. Environmental Protection Agency, Research Triangle Park, North Carolina 27711, EPA/600/3-85-035, 1982.

MATHUR R. et al. Multiscale Air Quality Simulation Platform (MAQSIP): initial applications and performance for tropospheric ozone and particulate matter. *J. Geophys. Res.*, 110(D13308), 2005. doi:10.1029/2004JD004918.

MENG, Z.; DABDUB, D.; SEINFELD, J. H. Size-resolved and chemically resolved model of atmospheric aerosol dynamics. *J. Geophys. Res.*, 103(D3), p.3419-35, 1998.

MORRIS, R. E. et al. Model sensitivity evaluation for organic carbon using two multi-pollutant air quality models that simulate regional haze in the southeastern United States. *Atmos. Environ.*, 40, p.4960-72, 2006.

PETERS, L. K. et al. The current state and future direction of Eulerian models in simulating the tropospheric chemistry and transport of trace species: a review. *Atmos. Environ.*, 29(2), p.189-222, 1995.

PUN, B. K. et al. Response of atmospheric particulate matter to changes in precursor emissions: a comparison of tree air quality models. *Environ. Sci. Technol.*, 42, p.831-7, 2008.

REYNOLDS, S. D.; ROTH, P. M.; SEINFELD, J. H. Mathematical modeling of photochemical air pollution I. formulation of the model. *Atmos. Environ.*, 7(11), p.1033-61, 1973.

RUSSELL, A. Regional photochemical air quality modeling: model formulations, history, and state of the science. *Annu. Rev. Energy Environ.*, 22, p.537-88, 1997.

RUSSELL, A.; DENNIS, R. NARSTO critical review of photochemical models and modeling. *Atmos. Environ.*, 34(12-14), p.2283-324, 2000.

SARWAR, G. et al. Impact of an updated carbon bond mechanism on predictions from the CMAQ modeling system: preliminary assessment. *J. Meteor. Climatol.*, 47, p.3-14, 2008.

SCHERE, K. L.; POSSIEL, N. C. U.S. EPA (Environmental Protection Agency) Regional Oxidant Model – background and overview, U.S. Environmental Protection Agency, Washington, D.C., 1984. EPA/600/D-84/155 (NTIS PB84201680).

SCHEFFE, R. D.; MORRIS, R. E. A review of the development and application of the Urban Airshed model. *Atmos. Environ.*, 27B(1), p.23-9, 1993.

SEIGNEUR, C. Current status of air quality modeling for particulate matter. *J. Air Waste Manage. Assoc.*, 51, p.1508-21, 2001.

SEIGNEUR, C. et al. Modeling atmospheric particulate matter. *Environ. Sci. Technol.*, 33, p.80A-84A, 1999.

STOCKWELL, W. R. A homogeneous gas phase mechanism for use in a regional acid deposition model. *Atmos. Environ.*, 20, p.1615-32, 1986.

WAYLAND, R. A. et al. Communicating real-time and forecasted air quality to the public: current state and future plans. *EM*, p.28-36, 2002.

WONG, D. C. et al. WRF-CMAQ two-way coupled system with aerosol feedback: software development and preliminary results. *Geosci. Model Dev.*, 5, p.299-312, 2012. doi:10.5194/gmd-5-299-2012.

YAMARTINO, R. J. et al. The CALGRID mesoscale photochemical grid model-I. model formulation. *Atmos. Environ.*, 26A(8), p.1493-512, 1992.

ZHANG, Y. Online coupled meteorology and chemistry models: history, current status, and outlook. *Atmos. Chem. Phys.*, 8, p.2895-932, 2008.

ZHANG, Y. et al. Application and evaluation of two air quality models for particulate matter for a Southeastern U.S. episode. *J. Air Waste Manage. Assoc.*, 54, p.1478-93, 2004.

22

Urban and Regional "Green" Mobility: Focus on Transportation

Nadja Lisboa da Silveira Guedes[1]

Introduction

Regional Transportation and Urban Mobility is a topic that is particularly inserted within the field of Architecture and Urbanism. Many countries, such as Brazil and China, are concerned with continuously increasing the productive population, trade and number of vehicles.

This theme is concerned with new concepts and discussions regarding urban mobility and sustainable transport ("green transportation"), according to conceptual social aspects.

Currently, social awareness regarding the need to live sustainably is growing. This, in addition to new paradigms, brings political pressure, so that each day more relevance to the topic is printed. There is a mobilization of governments in general, civil society groups and individuals, under the systemic vision of globalization.

Sustainability is revolutionizing the paradigms and issues related to mobility. Thus, there are new themes and concepts that incorporate economic, social and citizenship visions. The need to adopt policies,

1 Ufes, Vitória, Espírito Santo, Brazil. E-mail: *nadjaguedes@gmail.com*.

management plans and strategic planning can be observed, which avoids bottlenecks in transport and minimize environmental damage, including towns and cities in development.

- This article includes the following contents:
- Concepts of mobility;
- Modals of transportation;
- Resources and Techniques for Managing Mobility and Smart Technologies;
- Actions of Government, the Private Sector and Civil Society Organizations;
- Studies to improve urban mobility conducted researchers at Espírito Santo Federal University – UFES;
- The Potential for Future Research Projects.

China and Brazil are the focus of most examples given to discuss the content.

Travelling by bicycle in Shanghai (China) is a positive aspect for sustainable mobility that can be highlighted. However, there is an intensive flow of points of conflict regarding traffic, which makes it hard to safely manage mobility.

Other mobility problem is detected when the user does not see the necessary information in time. This happened during the technical visit to Minhang Campus in Shanghai (Figure 1).

Figure 1. Bus B of TopChina at Minhang Campus, JTU
Picture taken in July, 2011.

Bus B, while transporting the TopChina participants, was taking them to the library and had to go under a small overpass. However, the bus had to stop before this point because there was a doubt concerning the height limit. This problem could be solved with advanced information captured by electronic devices or signal plates.

We intend to explore the discussion of emerging themes regarding cities and life quality that are deeply inserted into environmental issues. One of the best ways to find solutions is to interchange experiences and learn from one another and reporting experiences – these are the learned lessons.

Given the above, the complexity of urban mobility versus transportation is clear. Today, a day's mobility conditions involve problems, such as: congestion, noise and air pollution, traffic-related accidents, environmental costs, social costs, delays on transportation, which can mean loss of production time, in its broadest meaning.

Objectives

The contents attune to the goals outlined.

a) General Purpose

Address and discuss "green" Transportation for good Urban and Regional Mobility. This purpose presupposes a consideration for preservation and well-being in cities/regions, and highlights the potential of Architecture and Urbanism and the importance of interdisciplinary knowledge.

(b) Specific Objectives

Some specific goals were developed, the purpose of which was to:

- List the Transport Systems and Modals, pointing out traffic issues, service Level, Safety and Environment.
- Conceptualize: Urban and Regional Mobility "green" and accessibility.
- Expose the state of Art and Resource Management Techniques for Mobility and Smart Technologies.

- Show the importance of the Government role, including Brazilian Programs, Federal Government, Municipalities and the Espírito Santo – Brazilian state.
- Address Environmental Impacts and Mitigation Ventures relating to transportation.
- Highlight the benefits of applying new techniques and technologies and the potential to enact Projects and Research in the area of Architecture and Urbanism.

Contribution

This study intends to contribute towards awareness regarding the field and role of transportation. In addition, this study hopes to assist Architects and Urban planners in their task to provide quality working under a systemic approach of "green" Mobility, as well as to encourage the rescue and/or to transform concepts and values, providing experience exchanges, critical learning to educate this context of Architecture and Urbanism, with a vision of social welfare, development, life and environmental preservation.

Questions and Subjects

Some questions were posed in order to achieve these objectives and provide guidance for reflection.

- Does the environmental improvement in the mobility in cities and regions result in a better world to live in it?
- What do you think should be done in the city you live in?
- Can a transportation area allow you exercise your "green" citizenship?
- Does high-tech help to achieve this?
- Are we in time?

These questions lead to the investigation of some related subjects: mobility concept; transport systems; urban mobility management, technologies, transport in Brazilian cities such as Vitória.

Sustainable Mobility

This topic contains a reference to a document entitled Mobility 2030: Meeting the Challenges to Sustainability (WBCSD, 2003), which includes data and statements, such as:

- Reduce conventional emissions (ex. Low- carbon Emission, as dioxide and carbon monoxide) from transport so that they do not constitute a significant public health concern anywhere in the world.
- Technology could drive conventional emissions down in developing countries.
- Beyond Copenhagen, the WBCSD focuses on the continued need for business to help design the institutions, mechanisms, innovative technologies, solutions and tools needed to deliver emissions reductions – Low-carbon Emissions.

São Paulo and Shanghai Mobility

This project studied mobility in four world cities: Bangalore – India, Dar es Salaam –Tanzania, São Paulo – Brazil and Shanghai – China (WBCSD, 2003).

"*China is now responsible for roughly a quarter (24%) of total global emissions and China's economic growth shows no signs of leveling off.*" São Paulo and Shanghai have similar problems as they did more than 60 years ago – increasing population. This means: special conditions of accessibility – in all modals – and special care.

The largest energy consumer in the world is the USA. China's rapid growth has turned it into the second largest. Since 1993, when China became a net importer of petroleum, its import dependency has risen every year, now reaching almost 52%. The International Energy Agency (IEA) projects that it will import up to 82% of its oil by 2030.

In developing world cities "*rapid growth, population density, poverty and inequality, limited public capacity and resource shortages add further to the challenge of enabling people and goods to move about sustainably*".

The main highlighted problems regarding transportation management are:

- Transport-related health and safety risks remain.
- The bad state of roads and vehicles.
- Poorly controlled intersections and inexperienced drivers are the main causes for transport-related deaths and injuries.
- Excessive number of undisciplined bicycle riders.
- Congestion is getting worse.
- High tech – e.g. Intelligent Transport Systems and Mechatronics – needs specialized jobs.
- International/domestic load transportation control and delays.

Some mobility management principles have been adopted in Latin America. The new concept of urban management is increasingly applied in its cities with some initiatives already underway in Brazil, Colombia, Chile, Mexico and Venezuela.

Nevertheless, there is lack of sufficient awareness so that comprehensive policies can be developed. Nevertheless, some measures aimed at public transportation have been taken, such as BRT – Bus Rapid Traffic – and BRS – Bus Rapid Speed. BRT and BRS are techniques that use information technologies and intelligent systems for transportation planning to satisfy the continuous increase in users' demand for public transportation, which operates in a preferential manner.

These factors show that managing the mobility is not a trivial problem. Thus, a Mobility Management concept is needed as a Transportation Planning technique.

Transportation Planning

This subsection includes definitions regarding Transportation Planning, as mentioned by Anselmo Neto & Guedes (2010). They highlight:

- Pereira et al. (2002), who state the concept is related to seeking transportation alternatives to reduce the negative effects in a population's daily life, encouraging the use of more sustainable forms of transportation such as public transportation, bicycling, traveling on foot and integration between different systems, using smaller volumes of financial resources.

- Real and Balassiano (2001) considers it to be a quite broad concept, incorporating a set of specific issues regarding the adequacy of existing handling of the different flows in a transport network. It is characterized in terms of democratic, flexible, economical and environmentally friendly solutions.
- Porto (2002) states that the basic principle is to create strategies for moving people and goods in a more balanced social, economic and environmental way, employing not only engineering techniques, but also tools to support the organizational field of information and telematics. As developed strategies, policies, application and objectives are different, they depend on the approach method and the needs of cities or places where the concept is adopted.
- Rocha et al. (2006) considers that it can be understood as the search for a more stable balance between the supply of transport infrastructure and the proper treatment of demand for travel (displacement); admits the possibility of different solutions in search of that balance, and considering the rationalization in the use of financial resources and ensuring the reduction of environmental impacts generated by transport systems.

The European Union has 40 (forty) projects for Mobile Management – MM, while the United States has Transportation Demand Management – TDM. These management models have been consolidating the principles which encourage the use of transport, which generate a lower environmental impact and road, thereby restricting the use of private cars and a way to better organize some of the cities' areas, the objectives being:

- to reduce congestion, accidents – traffic safety;
- to optimize travel times; and,
- to give importance to pedestrian needs.

Urban Mobility X Transportation X Environment

Firstly, today the three subjects are interconnected. It is necessary to know some transport classifications in order to understand and realize some of the issues' scale, scope and the transportation system's multidisciplinary approach. Some questions were formulated and related to different fields of science.

Some transport classifications and subjects or fields:

- Transportation by land, air, water, space (type of transportation related to ambient)
- On cables, rail, sea, river, lake, air, road, in vacuum, underground (means of transportation)?
- Do you go by aircraft, bicycle, train, GVT, metro, bus, lifter (modals of transport)?
- Public (available for all – taxi, metro…) or private transportation? Are the roads in suitable conditions for vehicles and pedestrian circulation? (Traffic and Civil Engineering, Education).
- Is there traffic congestion (Traffic, Urbanism)?
- Are the drivers stressed? (Psychology and Transportation)
- Is there a possibility for bacterial contamination in crowded environments, such as on buses, terminals or stations? (Medicine, Biology)
- How form should advertisements take on transportation vehicles? (Design, Marketing, Communication, Planning)
- Do roads need intervention in their design? (Architecture and Urbanism, Transportation, Topography)
- Which are the public and private actors in the transportation process? (Transportation management)

Vitória Case

This section presents a Brazilian city that is in line with economic and population growth, demanding better urban mobility management.

Urban Mobility Policy for Vitoria City (ES, Brazil)

Vitoria, Espírito Santo, Brazil – Vitoria is a port city that works with imports and exports, mainly stones, ore and pulp in addition to being a major center of business attraction due to the gas, oil and oil produced and exploited by Petrobrás/ Transpetro (logistics operation).

There is an observed congestion problem at the ports and airport, which causes a reduction in capacity. The infrastructure problem is critical and is probably a political and financial issue.

Vitória has 325,453 habitants (IBGE, 2010 apud Anselmo Neto, 2010). The city acts as a passage for many residents who make the routes between Cariaca-Vitória and Vila Velha-Serra (Grande Vitória – Great Vitoria), causing an increase in the local (capital) traffic flow. Internal and external travels into Vitória show the following trend: internal from 70%, in 1998 to 86.2%, in 2007; external from 30%, in 1998, to 13.8%, in 2007.

Data from the Economic and Applied Research Institute (Ipea) show that 44% of Brazil's population use public transportation. Other data show that in Espírito Santo (ES, Brazil), in 2000, there were 10.3 persons for each car, in 2010, there were 5.5 persons. There are 660,000 cars out of a total of 1,300,000 vehicles (Ipea apud Melo, 2011).

The modal division data show that there is predominance in the following order: 29% driver; 22.9% city bus; 21.4% on foot; 10.6% car passenger; 5.7% intercity bus; 2.7% bicycle moped; 1.5% motorcycle; 1.3% school transport (PDTMU – Transportation Master Plan and Urban Mobility, 2010).

Urban Mobility within urban perimeter presents several problems by: increasing the road network that did not follow the growth of the fleet; the Vitória island's conformation and the local topography working as a barrier to the expansion of urban road or railway networks.

Vitória Projects

The focuses of the projects are BRT – Bus Rapid Transit, Train (light rail), and restructuring of the city's main roads and construction of loops and overpasses, urban studies and actions – improving mobility and intermodal transportation. Encouraging public transportation such as buses, trains, subways, ferries. Moreover, the purpose of the projects was to improve pedestrian routes thereby increasing their use.

The use of Intelligent transport systems is considered, because technologies such as Telematics (communication + computers) and information for processing and communication, sensing, navigation and control technology are applied to improve the management and operation of transportation systems, these with compatibility and interoperability.

Figure 2 shows a proposal for ITS technology application for urban congestion, in Vitória City, to avoid congestion and preserve the environment.

Figure 2. Proposal for urban congestion charging.

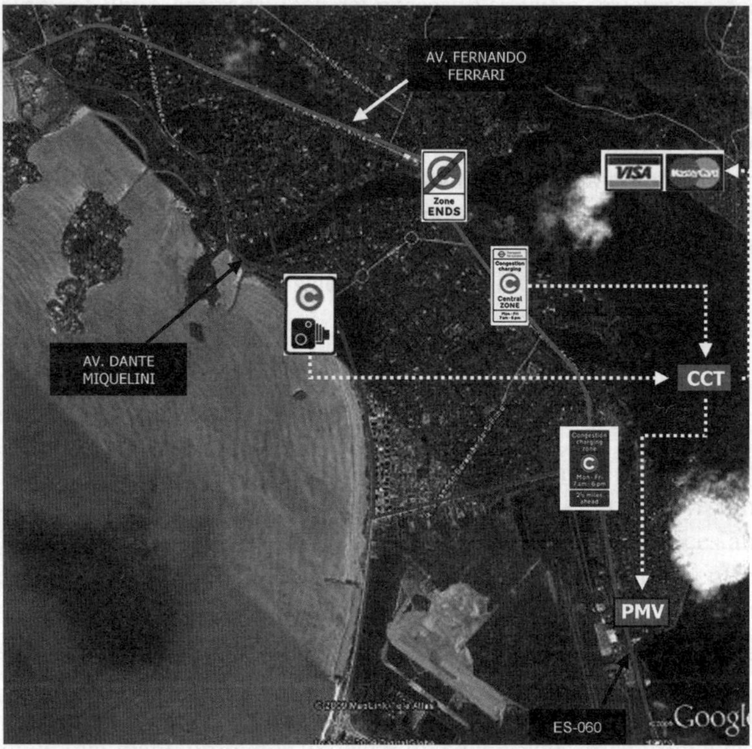

Proposals are the use of cameras, variable message signs that indicate alternative routes and charge through reading/write chips, credit cards and electronic communication with the control center. Figure 3 shows the use of a variable message panel – ITS technology that signals underground parking.

General Considerations

In addition to the aforementioned methods to solve urban circulation problems, other initiatives such as traffic calming measures, geometrical and physical interventions have been developed in order to focus on low speeds and cleaner forms of displacement, particularly pedestrians and bicycles within the European slowdown philosophy of life.

Figure 3. At the bottom there is an example of variable message paneling in Shanghai, in July, 2011.

There is a potential development of research and applications in Architecture and Urbanism, which mainly involve Transport, Modern techniques and information technologies/communication. These have modified the inter-relationships and the "frontiers of knowledge" which trend to be extinct, setting up a new conceptual paradigm for Urban Mobility that focuses on social and environmental benefits, including all modalities of Transportation.

Acknowledgements

We would like to thank the sponsors of the International Mobility Program and the Undergraduate Students and Teachers – TOP CHINA 2011.

References

ANSELMO NETO, L.; GUEDES, N. L. S.; BERTOLDE, A. I. 2010. *Contribuição à avaliação do nível de serviço de vias considerando emissão de poluentes e tempo de produção*. Dissertação de MSc, PPGEC, UFES, Vitória (ES), Brasil.

ESPÍRITO SANTO. PDTMU – *Plano diretor de transporte e mobilidade Urbana do Município de Vitória*. Secretaria Municipal de Transportes e Companhia de Desenvolvimento de Vitória. Espírito Santo, Vitória: PMV-ES; CDV, Brasil, 2010.

______. *Plano Diretor de Transporte Urbano do Estado do Espírito Santo (PDTU)*. Secretaria de Transportes e Obras Públicas, Brasil, 2008.

UE. *Mobility and Transport Documents*. European Commission. Available at: *http://ec.europa.eu/transport/urban/urban_mobility/urban_mobility_en.htm*. 2011. Accessed in June, 2011

GUEDES, N. L. S. *Modelo conceitual de liberação eletrônica de contêineres em fronteiras internacionais*. Tese de D.Sc., PET, COPPE, UFRJ, Rio de Janeiro, Brasil, 2005.

MELO, V. Número de carros aumenta e causa transtorno no trânsito capixaba. *Folha Vitória*. Date: 24/01/2011.

SWECO ITS (org.). *Green and ITS*: an overview of innovations for sustainable transport system in Stockholm. Sundlyberg: Sweden, 2009.

WBCSD-World Business Council on Sustainable Development. 2003. Meeting the Challenges to Sustainability, *www.oecd.org/sd-roundtable/papersandpublications/39360485.pdf*, Accessed June 2011.

23

Conservation of Mangroves in Urban Areas

Manuel Antonio de Andrade Furtado-Neto[1]

Introduction

The value of mangrove ecosystems has been extensively discussed worldwide (Schwarz, 2003), because mangroves are unique ecosystems with characteristically complex plant communities that exist along sheltered coast line areas in the tropical and sub-tropical inter-tidal regions of the world, such as in bays, estuaries, lagoons and creeks (IUCN, 2006).

Mangroves sustain diverse flora and fauna species on a large scale and provide many ecosystem services such as coastal protection from storm, shoreline reduction and riverbank erosion, thereby stabilizing sediment deposition and absorbing pollutants. Mangroves also provide important nutrient inputs and act as primary sources of energy for tropical estuaries. The biomass productivity of mangroves is relatively high and they act as a reservoir in the assimilation of waste. Mangroves stabilize the shoreline and act as bulwark against the devastating impact of hurricanes and cyclones as well as encroach-

1 Instituto de Ciências do Mar e Departamento de Engenharia de Pesca – Universidade Federal do Ceará, Fortaleza, Ceará, Brasil. E-mail: *mfurtado@ufc.br*.

ment by the sea in addition to mitigating soil erosion (IUCN, 2006; Macintosh; Ashton, 2002; MEF, 2013).

The concept that mangrove extension is decreasing has been spread worldwide over the last few decades, which is mostly due to discussion regarding other uses, in particular urbanization, tourism development and aquaculture (ITTO, 2002).

However, regional environmental changes have resulted in an expansion of mangrove areas in many coastal sites, such as in Thailand, Bangladesh, Mozambique, Venezuela and New Zealand, because the erosion of upland soils has increased sedimentation in coastal areas (Aizpuru, 2000; Perez et al., 2002; Schwarz, 2003).

The rapid response of mangroves to regional and global environmental changes and their huge ecological importance for coastal ecology and for the millions of people dwelling along tropical coasts, makes the continuous mapping of this vegetation a fundamental tool for coastal management plans and for maintaining the sustainability of coastal areas (Maia; Monteiro; Lacerda, 2006).

This paper describes various aspects related to the conservation of mangroves in urban areas and presents data regarding the conservation of mangroves in Brazil and China.

Conservation Status of Mangroves in Gobal Scale

The total global total of mangroves at the beginning of the 21th century was just over 152,000 square kilometers. There has been a considerable reduction over original natural levels due to human use and conversion. Accurate means of determining original mangroves cover are not available, but this cover was certainly greater than 200,000 square kilometers. However, even before these losses, they were a rare global habitat (Spalding; Kainuma; Collins, 2010). According to the FAO (2007), mangroves constitute less than one per cent (1%) of tropical forest worldwide, and less than 0.4 % of the total global forest.

The global distribution area of mangroves was approximately 188,000 square kilometers in 1980. Twenty-five years later, in 2005, this area had been reduced to 152,000 square kilometers. The total reduction of 36,000 square kilometers is equivalent to the size of

Taiwan. The conversion to aquaculture lands is the main reason for this reduction, accounting for an estimated 52% loss in mangrove areas. Deforestation is responsible for 26% of this reduction, while the remaining is caused by coastal development (FAO, 2007; MFA, 2012).

Mangroves are widespread within tropical latitudes and show variation in total area coverage. There is a high abundance of mangroves found on the wetter coastlines of South and Central America, West and Central Africa, and in a broad sweep from North-East India, through Insular South-East Asia, into Northern Australia (MFA, 2012).

In China, the northernmost boundary of naturally occurring mangrove has an approximate total area of 13,646 hectares. Mangrove growth is comparatively slower and trees are usually smaller (less than four meters). Furthermore, mangrove forests have hardly been used by Chinese people as lumber or similar wood construction. The destruction of mangrove in China is mainly attributed to the rampant felling of mangrove trees for the purpose of wood fuel (UNEP/GEF, 2003). Mangroves have been placed at the top of the list for protected resources in China, appreciating its significant indirect ecological value, according to the China Mangrove Protection, Management, and Utilization Plan (Weidong et al., 2000).

International pressure has been increasing to conserve and restore mangroves due to the global attention placed on sustainable resource use and biodiversity conservation. This pressure has allowed concrete actions to be taken globally.

According to the Ministry of Foreign Affairs, Republic of China-Taiwan (MFA, 2012), there are three main objectives in the global restoration of mangrove ecosystem:

a) Conserving the Natural Ecosystems and Landscapes: The planning and construction of conservation area is intended to maintain the ecology's current condition. The prevention of human development and damage is also the most common method adopted for environmental conservation.
b) Conserving the Sustainability of Natural Resource Production: The main objective of restoration is for the land to recover its ability to produce in the future.
c) Protecting Coastal Areas: Apart from focusing on conservation areas and sustainable conservation, the cultivation of mangroves

is also undertaken to restore the mangrove forests that were damaged by natural hazards.

Conservation of Mangroves in Brazil

Managing mangroves in rural and urban areas is a challenge. According to Spalding, Kainuma and Collins (2010), the oldest known legal document that offered partial protection to mangroves was a Decree signed in 1760 by King José I of Portugal. The decree was designed to ensure a sustained supply of mangrove bark for the tanning industry in parts of Brazil, stating that mangroves could not be cut for other uses unless their bark was utilized for its tannins, ensuring that timber or firewood was taken from other less valuable species.

Brazilian mangroves are present along 7,408 km of the country's coastline, covering approximately 25,000 square kilometers (km^2). Only one coastal state in the country does not have mangrove coverage, namely the state of Rio Grande do Sul, in the extreme south of Brazil. Mangroves exist from the border of Amapá State with French Guiana, just above the Equator (04°30′N), to beyond the Tropic of Capricorn (28°30′S), near Laguna, Santa Catarina State (Shaeffer-Novelli et al., 2000).

According to Magris and Barreto (2010), many studies related to the distribution, structure and variability of mangrove areas in Brazil have been conducted over the last 25 years, which has generated a large amount of information (Bernini; Rezende, 2004; Bernini; Rezende, 2010; Cavalcanti et al., 2009; Cunha-Lignon et al., 2009; Menezes; Berger; Worbes, 2003, Menezes; Berger; Mehlig, 2008; Schaeffer-Novelli, 1989; Schaeffer-Novelli; Cintrón-Molero; Adaima, 1990; Schaeffer-Novelli; Cintrón-Molero, 1999; Silva; Bernini; Carmo, 2005; Soares; Schaeffer-Novelli, 2005; Vedel et al., 2006).

In a recent study, Magris and Barreto (2010) mapped mangrove habitat and assessed the protection of this environment across the coastal protected areas using Landsat satellite images integrated with a Geographic Information System (GIS) along the entire Brazilian coast. According to their data, a total of 1,071 million of hectares of mangrove forest was registered in Brazil, with 86% of this value present along the macrotidal coast.

According to Maia et al. (2006), a large majority of the mangroves in Brazil (94%) are concentrated in the North (34%) and Northeastern (60%) regions. The other two geographic regions, South and Southeast, together only account for 6% of the mangroves areas (Figures 1 and 2).

Figure 1. Map of Brazil showing geographic Regions: North; Northeast; South; Southeast; Central-west. Source: original map drawn by author.

In Brazil, mangrove habitats have a high level of protection with almost 83% of the area of mangrove cover being located within protected areas when three levels of governance are considered – municipal, state, and federal. A total of 77% of protected mangroves are situated in protected areas of sustainable use.

Figure 2. Distribution of mangroves area (%) by geographic Regions in Brazil. Source: Maia et al. (2006).

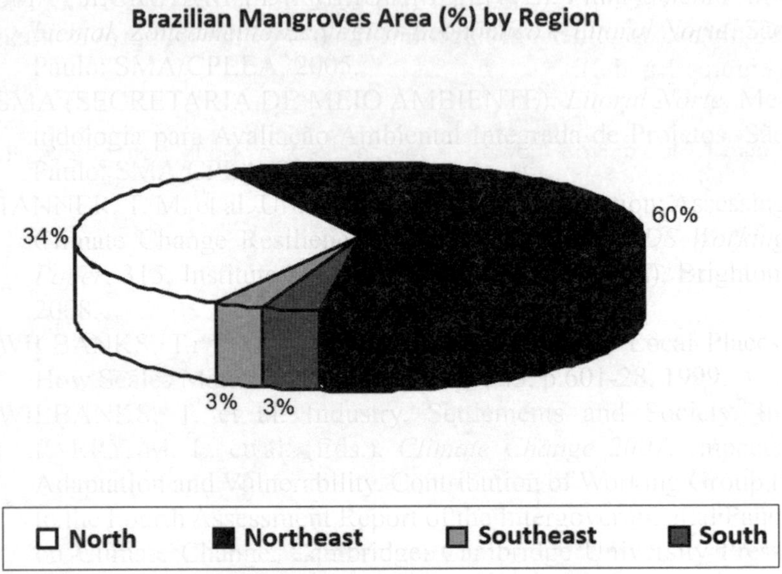

Figure 3 shows that the total area of mangroves grew from 1978 to 2003 in five Brazilian States: Piauí, Ceará, Rio Grande do Norte, Paraíba and Pernambuco, according to Maia et al. (2006).

Mangrove areas expanded in these states from 1978 to 2003 according to data from Herz (1991), while overall expansion reached about 37%. The largest expansion occurred in Pernambuco state (+67%) with the smallest taking place in Rio Grande do Norte (+19.9%). The fastest mangrove expansion occurred between 1978 and 1990. Most of the expansion during this period occurred in abandoned salt pans. Salt production, a major economic activity in the region during most of the 20th century, was drastically reduced by the 1970's; today being restricted to the Rio Grande do Norte (Maia; Monteiro; Lacerda, 2006).

From 1978 to 2003, mangrove extension in Northeastern Brazil occurred mainly as a result of saline intrusion and changes in sediment deposition along river margins, as can easily be seen by the different generations of mangrove growth along many estuaries on the mapped area. These changes, which occurred in the estuarine morphodynamics, were mostly due to the increase in rivers damming and withdrawn freshwater due to the expansion in irrigated agriculture and coastal human population over the last two decades of the 20th century (Maia; Monteiro; Lacerda, 2006). Urban areas of cities with urban areas home to more than 1,000,000 of inhabitants, such as São Luis and Fortaleza in Northeastern Brazil, are included in these numbers.

Figure 3. Total mangroves growing area in 5 different States of Northaestern Brazil. Source: Maia et al. (2006).

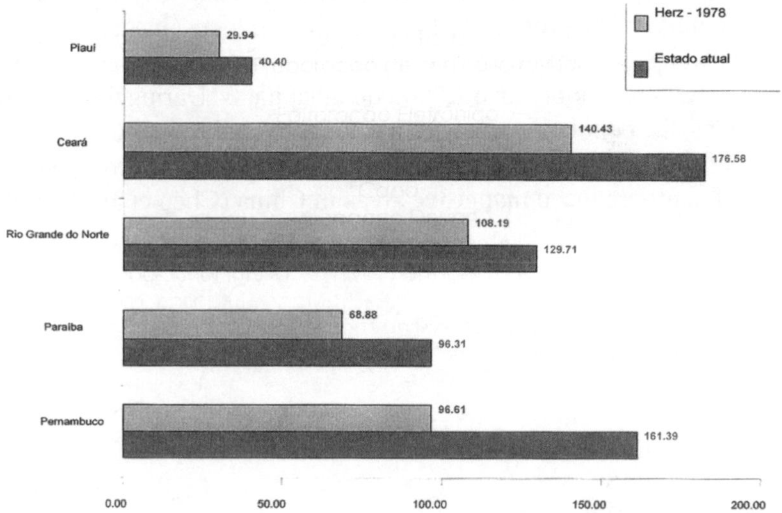

Conservation of Mangroves in China

Chinese scientists have conducted many research projects on mangroves since the 1950s, including one review on Chinese mangrove research conducted by Li and Lee (1997). Since the 1990s, investments by government have improved research on mangroves in China. As a result of these investments, a series of books regarding Chinese mangroves were published during the last decade (Chen; Miao, 2000; Fan, 2000; Lu; Ye, 2006; Tam; Wong, 2000; Wang; Liao; Wang, 2002; Wang; Wang, 2007). However, more than three quarters of scientific papers and most books were written in Chinese, which are not accessible to the international research community.

The Chinese government has launched a series of programs to protect mangroves since 1980s and has established mangrove ecosystems as high-priority areas for improving environmental and living resource management. During the last three decades, a total number of 34 natural mangrove conservation areas have been established, which accounts for 80% of the total existing mangroves areas in China. Mangrove restoration areas in Mainland China accounted for less 7% of the total mangroves areas in 2002 (Chen et al., 2009).

In China, mangroves are naturally distributed in the Southeast of the country, in the provinces of Hainan, Guangdong, Guangxi, Fujian, Hong Kong, Macao and Taiwan (Figure 4), which have at least twenty urban areas with more than 1,000,000 inhabitants. During the last few decades, the population has boomed and the rapid economic development in agriculture, aquaculture, industry and urban construction has significantly reduced mangrove areas in China (Chen et al., 2009).

Figure 4. Map of China showing the Provinces of Hainan, Guangxi, Guangdong, Fujian, Zhejiang, Hong Kong (H), Macao (M) and Taiwan – provinces with surface areas of existing mangroves in China. Source: original map drawn by author.

According to the latest survey by the State Forestry Administration in 2001, the total mangrove area in mainland China was 22,025 ha, which was only 44% of what existed in the 1950s (SFA, 2002). The existing mangrove area in Hong Kong, Macao and Taiwan was 727 ha in total according to Chen (1997), Leung (1998), and Hsueh and Lee (2000). Thus, the total area of existing mangroves in China was approximately 22,752 ha (Table 1).

Table 1. Surface areas of mangroves in China by province, according to SFA (2002);[1] Chen (1997);[2] Leung (1998)[3]; Hsueh and Lee (2000),[4] and reforestation area up to 2002 (Chen et al., 2009).[5]

Location	Total mangrove area (ha) 2002	Reforestation area (ha) Up to 2002
Hainan Province	3 930[1]	60[5]
Guangxi Province	9 084[1]	672[5]
Guangdong Province	8 375[1]	1 093[5]
Fujian Province	615[1]	596[5]
Zhejiang Province	21[1]	257[5]
Hong Kong	380[2]	-
Macao	60[3]	-
Taiwan	287[4]	-
Total	22 752	2 678[5]

According to Chen et al. (2009), many efforts and achievements have been made regarding mangrove restoration in China, but there is still a lack of a universal standard system for evaluating such efforts and achievements. Collaborations among Chinese governmental agencies (such as State Administration of Forestry), research institutions and local communities are strongly encouraged to establish such an evaluation standard system so that mangrove restoration can be improved.

References

AIZPURU, M. *Global assessment of cover change of the mangrove using satellite technology at medium to high resolution.* Ispra (Italy): EEC Jonit Research Center, 2000.

BERNINI, E.; REZENDE, C. E. Variação estrutural em florestas de mangue do estuário do rio Itabapoana, ES-RJ. *Biotemas*, 23(1), p.49-60, 2010.

BERNINI, E.; REZENDE, C. E. Estrutura da vegetação em florestas de mangue do estuário do rio Paraíba do Sul, Estado do Rio de Janeiro, Brasil. *Acta Botanica Brasílica*, 18(3), p.491-502, 2004.

CAVALCANTI, V. F. et al. Evaluating mangrove conservation through the analysis of forest structure data. *Journal of Coastal Research*, 56, p.390-4, 2009.

CHEN, S. P. The mangrove of Hong Kong. *Trop Geogr.*, 17, p.184-90, 1997.

CHEN, G. Z.; MIAO, S. Y. *Mangrove Species Kandelia candel and Wetland Ecosystem Researches*. Guangzhou: Sun Yat-Sen University Press, 2000.

CHEN, L. et al. Recent progresses inmangrove conservation, restoration and research in China. *Journal of Plant Ecology*, 2(2), p.45-54, 2009.

CUNHA-LIGNON, M. Estudos de Caso nos Manguezais do Estado de São Paulo (Brasil): Aplicação de Ferramentas com Diferentes Escalas Espaço-Temporais. *Revista de Gestão Costeira Integrada*, 9(1), p.79-91, 2009.

FAO. Food and Agriculture Organization of United Nations. Mangroves of South America 1980-2005. *Country Report, Forest Resources Assessment Working Paper*, n.139, Rome, 2007. 321p.

FAN, H. Q. *Mangroves*: The Environmental Guard of Coastline. Nanning: Guangxi Science and Technology Press, 2000.

HERZ, R. *Manguezais do Brasil*. São Paulo: Instituto Oceanográfico, Univ. São Paulo, 1991. 227 p.

HSUEH, M. L.; LEE, H. H. Diversity and distribution of the mangrove forests in Taiwan. *Wetl Ecol Manag*, 8, p.233-42, 2000.

IUCN. Conservation Benefits of Mangroves. The World Conservation Union – IUCN, Sri Lanka. *IUCN Policy Brief*, 10, p.1-6, 2006.

LEUNG, V. The distribution pattern of mangrove plant populations and its species composition in Macau. *Ecol Sci*, 17, p.25-31, 1998.

ITTO. *Mangrove Work Plan*. International Tropical Timber Organization – ITTO, Yokohama, Japan, 2002. 231p.

LI, M. S.; LEE, S. Y. Mangroves of China: a brief review. *For Ecol Manag*, 96, p.241-59, 1997.

LU, C. Y.; YE, Y. *Wetland Ecology and Engineering*. Mangroves as Example. Xiamen: Xiamen University Press, 2006.

MACINTOSH, D. J.; ASHTON, E. C. *A Review of Mangrove Biodiversity Conservation and Management*. Denmark: Centre for Tropical Ecosystems Research, University of Aarhus, 2002. 71p.

MAGRIS, R. A.; BARRETO, R. Mapping and assessment of protection of mangrove habitats in Brazil. *Pan-American Journal of Aquatic Sciences*, 5(4), p.546-56, 2010.

MAIA. L. P. et al. *Atlas dos manguezais do nordeste do Brasil*. Fortaleza: SEMACE, 2006. v.1. 125p.

MAIA, L. P.; MONTEIRO, L. H. U.; LACERDA, L. D. Changes in mangrove extension along the Northeastern Brazilian coast (1978-2003). ISME. *Glomis Electronic Journal*, v.5, p.1-5, 2006.

MEF. *Conservation and Management of Mangroves*. New Deli: The Ministry of Environment and Forest of India – MEF, 2013. p.1-3.

MENEZES, M., BERGER, U.; WORBES, M. Annual growth rings and long-term growth patterns of mangrove trees from the Bragança peninsula, North Brazil. *Wetlands Ecology and Management*, 11, p.233-42, 2003.

MENEZES, M. P. M.; BERGER, U.; MEHLIG, U. Mangrove vegetation in Amazonia: a review of studies from the coast of Pará and Maranhão States, North Brazil. *Acta Amazonica*, 38(3), p.403-20, 2008.

MFA. *The ecology of Mangroves in Taiwan*. Ministry of Foreign Affairs, Republic of China-Taiwan – MFA. 2012. p.1-15.

PEREZ, A. C. et al. Monitoring a recent delta formation in a tropical coastal wetland using remote sensing and GIS. Case study: Guapo River Delta, Venezuela. *Environment, Development & Sustainability*, 4, p.201-19, 2002.

SCHAEFFER-NOVELLI, Y. Perfil dos Ecossistemas litorâneas brasileiros, com especial ênfase sobre o ecossistema manguezal. *Publicação Especial do Instituto Oceanográfico de São Paulo*, 7, p.1-16, 1989.

SCHAEFFER-NOVELLI, Y.; CINTRÓN-MOLERO, G. Brazilian mangroves: a historical ecology. *Jornal Ciência e Cultura*, 51, p.274-86, 1999.

SCHAEFFER-NOVELLI, Y.; CINTRÓN-MOLERO, G.; ADAIMA, R. R. Variability of mangrove ecosystems along the Brazilian coast. *Estuaries*, 13, p.204-18, 1990.

SCHAEFFER-NOVELLI, Y. et al. Brazilian mangroves. *Aquatic Ecosystem Health and Management*, 3(4), p.561-70, 2000.

SCHWARZ, A. M. Spreading mangroves: a New Zealand phenomenon or a global trend? *Water & Atmosphere*, 11(1), p.8-10, 2003.

SFA. 2002. *Report of Mangroves Survey in China*. State Forestry Administration of China, 145p.

SILVA, M. A. B.; BERNINI, E.; CARMO, T. M. S. Características estruturais de bosques de mangue do estuário do rio São Mateus, ES, Brasil. *Acta Botanica Brasilica*, 19(3), p.465-71, 2005.

SOARES, M. L. G.; SCHAEFFER-NOVELLI, Y. Above-groud biomass of mangrove species. I. Analysis of models. *Estuarine, Coastal and Shelf Science*, 65, p.1-18, 2005.

SPALDING, M.; KAINUMA; M.; COLLINS, L. *World Atlas of Mangroves*. Okinawa: International Tropical Timber Organization (ITTO), 2010. 319p.

TAM, N. F. Y.; WONG Y. S. *Hong Kong Mangroves*. Hong Kong: City University of Hong Kong Press, 2000.

UNEP/GEF. *Final Report of Review of China National Mangrove Economic Values*. United Nations Environment Programme and Global Environment Facility – UNEP/GEF, Guangxi Mangrove Research Center, 2003. 11p.

VEDEL, V. et al. Holocene mangrove dynamics and sea-level changes in northern Brazil, inferences from the Taperebal core in northeastern Pará State. *Vegetation History and Archaebotany*, 15, p.115-23, 2006.

WANG, B. S.; LIAO, B. W.; WANG, Y. J. *Mangrove Forest Ecosystem and Its Sustaninable Development in Shenzhen Bay*. Beijing, China: Science Press, 2002.

WANG, W. Q.; WANG, M. *The Mangroves of China*. Beijing: Science Press, 2007.

WEIDONG, H. et al. The ecological values of mangrove ecosystems in China. *Ecologic Science*, 19(1), p.40-6, 2000.

24

Risk, Modernity and Networks: The Risk of Modernity and the Emergence of Networks in Industry

Luciano Modes[1]

Introduction

The background in which this analysis must be situated comprises a set of changes related to what here will be called modernity. This expression was created during the seventeenth century and is associated with a set of major changes that have extended the consequences to various points of the planet. With this, we identify modernity which, according to Anthony Giddens (1991), must be understood in a logic that breaks from the traditional:

> Modern social institutions are in some respects unique – distinct in form from all types of traditional order. Capturing the nature of the discontinuities involved, I shall argue, is a necessary preliminary to analyzing what modernity actually is, as well as diagnosing its consequences for us in the present day. (p.9-10)

Modernity opens new possibilities and allows for a fundamental reconfiguration of what we understand by *individual, family,* or even, *State*. For organizations, the new context also implies a series of

1 Universidade de Cuiabá – UNIC, Cuiabá, Mato Grosso, Brasil. E-mail: *lucianomodes@gmail.com*.

discontinuities and instabilities. In a way, we can say that modernity brought with it insecurity, both to the individual and by extension to organizations. The risk analysis of this set of contemporary changes has been shown to be one of the most promising fields for new approaches in Social Sciences, which are led, among others, by researchers such as Beck (1997; 1998) and Giddens (1991).

Commonly, insecurity has always been close to human beings: the tribes of hunters ran the risk of food scarcity due to sudden climatic changes; the medieval man was in danger of being ravaged by the forces of nature. It was clear at the dawn of modernity that it was imperative to understand the laws of nature in order to adapt to them (Bacon, 1984). We see, however, in the period that followed, a position adopted by humanity, not only adapting to nature but also reconstructing it in order to generate more security in this world of "so irrational". Forces Modernity is therefore, associated, *roughly,* with a period of rationalization and control of the forces of nature and risk reduction through technology.

It should be pointed out, however, that the technological advances that allowed old fears to be controlled, concomitantly, provoked the emergence of new problems, such as a side effect. In this sense, there can be at least two types of risk:

(i) For the first time in history, we are able to control of the elements of nature to avoid not only unprecedented disasters but also to create them, such as atomic wars;

(ii) The insecurity in modernity is also associated with the emergence of the expert systems, since we are totally at the mercy of the technologies and the technicians that master them.

In other words, in this era we have many resources at hand; however, we do not master the workings of the technologies.

Applying this context of fast and deep changes, which are more noticeable for the discontinuity and less for the maintenance of the established order, we ask ourselves: in relation to organizations, how is it possible to take a stance in this period? As previously mentioned, modernity hides the risk *within* its own proposed solutions. So: what are the risk factors associated with the survival prospects of organizations in modernity?

In order to analyze this case, we will focus on the automobile industry, mainly the American one. Overall, we have adopted the presupposition that the automobile industry develops networks in an attempt to share the risks and very high investments. This behavior generated by modernity (of being organized in the form of networks) serves as an example to show:

(i) The nature of the networks in modernity;
(ii) The prevalent place of the technologies that the institutions develop in their networks;
(ii) The centrality of the consumers as actors that, if sometimes silent, are those who determine the future of such networks.

Essentially, these three topics provide us with a general view of how the networks exist in modernity, their scope and precariousness. Yet, more specific questions arise: what are the characteristics of these networks created around the automobile industry? How were they formed? We will now briefly discuss these topics.

Companies in the logic networks: the intersectoral flows

Firstly, it is important to highlight the relevance of industry in the wider economy and, in order to narrow our focus, we will turn to the automobile industry. According to David C. Mowery and Nathan Rosenberg (2005),

> (...) the car – even if we ignore its contributions for the development and spread of mass production – was a unique transformative innovation. It brought with itself drastic alterations in the pattern of soil use. It completely changed the rhythm of urban life, including the spatial organization of work and residence, the patterns of socialization, leisure and shopping and led to the vast expansion of the suburbs. (p.62)

Even if it is possible to question the innovative capacity of the American automobile industry (Mowery; Rosenberg, 2005), its importance lies in the intersectoral flows that converge to it. Given the dimensions, scale of production and the market, this industry has become a spiral, to which a long chain of suppliers and service providers are "tied". These professionals will be responsible for the execution of a major part of the project created by the automobile industry in the supply of parts, materials and services. The colossal networks created around this industry are answers to uncertainty and changes in markets.

If in the beginning of the introduction of the Automotive Regime it was still possible to maintain a degree of relative autonomy, the Japanese example showed that the creation and maintenance of strong networks became vital in this market niche. The competence of the actors involved in the network (or their absence) will determine the future of the networks. Among them, the risks and also the possibilities are shared A parts supplier needs to be convinced of the anticipated success of the car project to be produced and make a bet with the automaker.

However, the success of a project will not be automatically guaranteed. It is exactly this uncertainty that moves the automobile sector towards the formation of networks. Within these networks organized by the large industries, the actors play a role determined by the automaker, in order to ensure as much success as possible. However, the suppliers are not the only actors crucial to the success of the car manufacturers. This is what the next topic will show.

Technological trajectories: the place of innovation and the course of technology in the automobile industry

To understand the nature of the networks created around the major automobile industries, it is necessary to analyze the place of technology in today's society. The trajectory of the technologies has escaped the control of the institutions, which provides a competitive environment. If one day it was under the control of big enterprises, nowadays, technology must be seen as an actor that composes the network. Its cooperation is not automatic, in view of the purposes of the most important actors, from which the networks are created.

Thus we could say that the greatest innovations used in cars appeared in other industrial sectors, but that over the years, they were transferred to the automobiles. However, technology, in the society we live in, seems to be more than a mere instrument used to determine the success or failure of certain organizations. We can mention the technologies as actors as we move away from the functionalist view and analyze its *trajectory in the modern societies* – where the technologies act in a very unexpected way to their creators.

Good examples of how the technology can become an actor in the network in which it was created can be researched in the American industry. *Xerox,* synonymous with photocopy, not only lost market inside a niche that it practically invented, but also saw its technology boosting other enterprises, which almost took it to bankruptcy. The Xerox technology was not only in the service of competitors, but also strengthened them.

Perhaps the most significant example of why the technologies should not be seen in a functionalist way in modern societies, but as actors that behave in ways to escape the control and manipulation, is the *Internet.* Born and commissioned by the United States Department of Defense as a military technology (Mowery; Rosenberg, 2005), the Internet, today, escaped the control of its creators. Although funded by the Department of Defense, it is still possible to find sites in the worldwide web promoting terrorism against the country that invented it or even hoarding new suicides bombers to participate in attacks. We are not saying that the American government might not want to stop it, but, that simply; this option was not available in many cases.

In the logic of networks proposed by Callon (1986a, 1986b), technologies emerge as actors since they act in relation to other actors, playing roles and can rebel against the actor-network that attempts to impose a script or a role on them. Their trajectories allow us to glimpse, in modernity, that they escape the simplistic or reductionist models. It is up to the organizations not only to develop technologies to be launched on the market, but also follow technological advances, subject to becoming obsolete by the very technology they set in motion.

The American industry, a leading developer of the automobile industry or, at least, its main consumer, now finds itself facing huge challenges generated by the technologies that they set in motion.

Even if our aim is not to track all the elements that seem to be leading the United States industry to share center stage of automobile production with several other competitors, it seems that one of the most relevant reasons is the insufficient capacity of this industry to keep up with technological innovations (Mowery; Rosenberg, 2005). Even if automobile technology has been developed in the USA as it has anywhere else in the world, it does not seem to be under the control of the Americans. Rather, the networks *within* are increasingly more complex, technology shows a private path, surprising the world when projecting small networks or when putting big actors on the market that have not adapted to the new times.

The networks created around the "spiral" of large organizations to which a long chain of suppliers is tied and where the technologies develop, are not complete in themselves. And also, the technological trajectories that we have mentioned do not occur in a vacuum. The *locus* in which the big networks are constructed, involving industry and suppliers and where the technologies run their trajectories, is the social environment and, in this space, the subjects must be taken into account as well. So, another actor appears that is fundamental to the success or failure of the networks; the subject.

The consumer has the strength: the empowerment of the market and the networks

In the logic of the modern society an actor that will decide the future of the networks emerges: the regular user of the technology, the one who will consume it. Consuming technology brings the interesting sense that it will be used, spent and consumed, which implies that it will be used to its limits, or, according to Kuhn: it will have its paradigm explored to the maximum, generating anomalies that will lead to improvements, in the scope of this paradigm, or even to its replacement for a new one (Kuhn, 2001).[2] Currently, this power is in the hands of the consumers as well as in the research and development laboratories of.

Modernity has brought with it a fundamental reconfiguration of power, as can be noted from the writings of Michel Foucault (1992). What the Japanese experience in the automobile industry has shown, through the reconfiguration of power in the Japanese Way, or modes of lean production, is that the adherence to new products is not so much pushed by industry, but driven by consumption. Nowadays, the approval of certain technologies or their ostracism is increasingly being

---

2 This idea was proposed by Dr. Joel Paese in the Sociology of Technology seminar promoted by the research group about science, technology and technological innovation of the UFMT.

determined by non-specialist consumers. They are micro-powers that emerge in a society that has capacity for decision.

The importance of the consumer or the power that is in his/her hands can be seen in the attempt of the automobile industry to produce models that follow the trends and tastes of those who will consume them. Every day, the consumer provides this list of innovations to the automakers. This can be illustrated in the following interview, mediated by a national automobile site with questions from consumers addressed to Honda at the launch of the *City* in Brazil.

> **Which measures will Honda adopt for the Civic to continue leading the sales in its segment?**
> The Civic has maintained a modern and bold design since 2006and remains one of the best cost/benefit models, not only in its segment, but also in the market. It brings the highest technology, as well as being a modern vehicle, with a large internal space and at a very competitive price. *It should continue to lead sales because of all these characteristics and also consumer choice.*
>
> **Doesn't Honda have any plans to enter the market of popular vehicles in Brazil?**
> At this time, the models on the market are in line with the wishes of the consumers of the brand. *In the future, the market will Honda tell whether or not it should enter this segment.*
>
> **Why did Honda not opt to launch the City LXL?**
> *A new version or engine model will be a market decision. If necessary, Honda will analyze and study the possibility.*[3]

Surely there is a complex relationship between producers and consumers, where each one concentrates a great deal of power but, according to what was presented, modernity brings to the hands of the final user, the "layman," a great deal of power: it is him/her who chooses from the set of available options, which technology he will use, spend, consume and so on.

Conclusion

Modernity has changed the social scenery deeply when it started on some of the basic elements of society, creating, among other things, new logics and social dynamics. These dynamics, as were highlighted

3 Available in: *http://carroonline.terra.com.br/index.asp?codc=3743*. Access in: August 23, 2009. (Emphasis added.)

at the beginning of this text, indicate a gloomy side of modernity, such as the increased risk to be faced and how to deal with it.

Overall, we could say that organizations have been urged to be reorganized within a logic of great risks. Hence, the formation of networks sharing the risks and provoking solutions that demand a lot of investments. In this logic, the emergence of these networks that oppose the idea of isolated industry must be understood as an answer to the new times.

However, as we attempted to highlight in this article, the networks are not closed in terms of the producers and suppliers that are tied to them. And it is this difference that makes them highly complex. The institutionalization of the big networks becomes more and more precarious as actors, outside the network control, grow in importance and power. A summary of the logic of the automobile industry indicated here can be seen in the following illustration:

Figure 1. The networks and their senses. Source: Author.

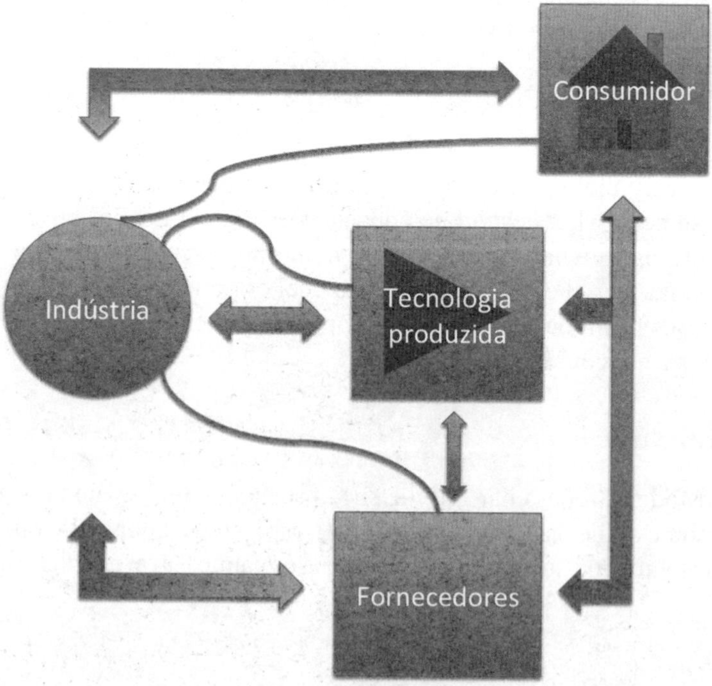

In this scheme the networks (the lines involving industry, suppliers, technologies produced and consumers) and the relationships involving these different actors (represented by arrows) are represented. It is important to emphasize that each actor relates to each other, without, necessarily, any mediation from the great automobile industry. This makes these networks much more open to the influences of each actor.

The presence of actors in a certain network is very unstable. This means that the devices that maintain adherence to a certain network may not be enough to maintain strong adhesion for a long time, unless they are periodically renewed. For example: consumers can freely exchange their products for others that they consider to be more convenient (as they often do).

The same precarious adherence by consumers to a certain brand can be noted in relation to technologies. Once a certain technology is launched in the market, it takes a trajectory based on the relationship with the consumers. In other words, the technology will be explored to its limit, which will generate the need to review the anomalies raised by this heavy use, or the weaknesses in the project, or even, to increase the right decisions. In any case, it is up to the industry to reappropriate the technologies launched by it, tweaking the boundaries of the technological paradigm. It is never too late to point out that, once a technology is put in movement, it will follow its course, permeating other networks that compete for the same space and triggering new solutions to the problems generated by the heavy usage of the consumers.

The technological trajectory proposed above allows us to understand how the companies that have launched revolutionary technology were themselves victims of the advances they have created. This happens because all the actors described relate to each other, generating possibilities and recombination, escaping from the control of the large industry, that is, even when tied around a network, the trajectories of each actor, through the combination with each other, escape the control of the actor who developed them.

We conclude, therefore, that the creation of networks is a fundamental part of any complex undertaking, such as in the production of cars; however, the nature of these networks is very precarious, to the point of transforming them into power structures that were unimaginable in pre-modern days. But, given its nature within the networks, anything can happen, including nothing.

References

BACON, F. *Novum organum ou verdadeiras indicações acerca da interpretação da natureza/Nova Atlântida.* Trad. José Aluysio Reis de Andrade. São Paulo: Nova Cultural, 1984.

BECK, U. *La sociedad del riesgo: hacia una nueva modernidad.* Trad. Jorge Navarro et al. Barcelona/Buenos Aires: Paidós, 1998.

______. A reinvenção da política: rumo a uma teoria da modernização reflexiva. In. GIDDENS, A.; BECK, U.; LASH, S. *Modernização reflexiva*: política, tradição e estética na ordem social moderna. Trad. Magda Lopes. São Paulo: Unesp, 1997.

CALLON, M. Some elements of a sociology of translation: domestication of the scallops and the fishermen of St Brieuc Bay. Originally published in: LAW, J. *Power, action and belief: a new sociology of knowledge?* London: Routledge, 1986a.

______. The sociology of an actor-network: the case of the electric ehicle. In: CALLON, M.; LAW, J.; RIPP, A. (Eds.). *Mapping the dynamics of science and technology: sociology of science in the real world. London: Macmillan,* 1986b. Available at: http://tecnologiaysociedad.uniandes.edu.co/2008/histecs200801/thesociologyofanactornetwork.pdf. Access in: August 24, 2009.

FOUCAULT, M. *Microfísica do poder*. Rio de Janeiro: Graal, 1992.

GIDDENS, A. *As consequências da modernidade.* São Paulo: Editora Unesp, 1991.

KUHN, T. S. A estrutura das revoluções científicas. São Paulo: Perspectiva, 2001.

MOWERY, D. C.; ROSENBERG, N. *Trajetórias da inovação: a mudança tecnológica nos Estados Unidos da América no século XX.* Campinas: Unicamp, 2005.

25

Urban conservation in Brazil: an overview

Leonardo Barci Castriota[1]

Introduction

During the rich experience of the TOP China Program, in which I had the opportunity to participate as a faculty representative from the Federal University of Minas Gerais, we were asked to develop a lecture based on a topic related to the issue of urban sustainability, which was the main topic of the program. I chose the theme of cultural heritage and urban conservation, which seemed very appropriate to explore the various issues related to sustainability.

Upon introducing the topic, I asked each of the students present – Chinese and Brazilians, to present cultural property that, in their minds, would represent the heritage of each country. The result was quite interesting: while the Brazilians chose a variety of items – ranging from historic buildings to aspects of intangible heritage such as the "capoeira" dance, the students from Peking University invariably listed items related to Chinese tradition and monumental buildings: traditional theater, the Forbidden City, among others.

---

1 Universidade Federal de Minas Gerais, Belo Horizonte, Minas Gerais, Brazil. E-mail: *leonardo.castriota@pq.cnpq.br*.

These answers were no surprise, as China has a long, rich history with an almost unbroken cultural tradition extending back some 5,000 years. However, this brief exercise struck me as indicative of the differences in *approach* between the two countries in relation to heritage preservation: while in Brazil, the European tradition of preservation emerged in the 1930s, China's seems to work within another tradition that has only recently begun to relate to the Western approach of historic preservation.[2]

The same differences are manifested in regards to urban conservation, which is a central theme in the current discussions on historic preservation, and is also an important topic that directly relates to the theme of urban sustainability, especially in countries like Brazil and China, that have experienced rapid economic growth. The Chinese case is particularly dramatic as Chinese cities have been undergoing major physical and socioeconomic transformations since the economic reforms initiated by the central government of China in 1978[3] and, as Whitehead and Gu pointed out, large-scale redevelopment has generated great pressure on conservation in many old-established towns and cities (Whitehead; Gu, TPR, 78 (5) 2007).[4]

In a country like Brazil, which is marked by an ideology of the new and whose landscape is extremely changeable, the preservation of entire ensembles of buildings from the past centuries, seems remarkable. The fact is that since the 1930s, against all odds, Brazil has adopted a conservation policy that, along with churches and important build-

---

2 As an example of this, we can notice that, despite its outstanding cultural legacy and a long tradition of conservation and restoration practice, China has had no national charter or set of guidelines regarding conservation and management of its cultural heritage. The first move in this direction was the issue of the so called "China Principles," a joint work by the Getty Conservation Institute and China ICOMOS in 2000.

3 More on this topic, see Arantes, 2011.

4 These authors also point out that in response to the destruction of historical areas in many Chinese cities and threats of further destruction, the preparation of conservation plans has become major undertaking of city management since the mid-1980s (Wang et al., 1999, p.10). "However, without a sound theoretical basis, the plan-making process has lacked firm foundations. As a consequence a great deal of inappropriate historical restoration and redevelopment has taken place. In seeking more effective strategies for urban conservation, it is important first to assess past and present research and practice and secondly to consider how methods and ideas developed elsewhere can with advantage be applied in China" (Whitehead; Gu, TPR, 78 (5) 2007).

ings, also protects urban complexes, through a mechanism known as "tombamento" (listing), which creates a system of co-responsibility between the owners of private property and the state. In order to try to understand this phenomenon, which is quite different from the Chinese case, I will present in this paper, albeit only briefly, the history of urban conservation in Brazil, focusing on its early days and trying to understand especially what led intellectuals and politicians to propose the preservation of complete ensembles as early as the 1930s.

Redescovering Brazilian Roots

Initially, it is necessary to draw attention to a peculiarity of the Brazilian case: heritage-preservation policies in Brazil have traditionally been drawn up and implemented by progressive intellectuals rather than conservative groups. One reason for this is that, in addition to its strong critique of traditional academic arts, the Brazilian modern movement has stressed cultural renovation and a search for its roots, while placing highly the issue of identity on the national agenda. As Helena Bomeny pointed out, modernist intellectuals in Brazil believed that "cosmopolitanism, immigration, foreignism, imitation, classicism's conventional and universalized formulas, timeless solutions, stilted language" were dangerous (Bomeny, 1994, p.77). Thus, while keeping in close contact with the European avant-garde, Brazilian modernists developed a peculiar relationship with tradition, refusing the idea of a radical rupture with the past. As summarized by the art critic Aracy Amaral:

> In the twenties, architecture, literature, and the visual arts were marked by a desire for formal renovation, making the early years of this century a time of cultural definition, particularly in São Paulo. The intellectual and art communities remained attentive to what was happening in Europe, especially in Paris, but emotional commitments lay with Brazil. Although educated in Europe, the Brazilian artists and writers who participated in the modernist movement – which had its historical beginning during the Week of Modern Art at São Paulo's Municipal Theater, 11 to 18 February 1922 – gradually came to realize that Brazilian reality was as important as formal renovation. (Amaral, 1995, p.14)

In this context, it should come as no surprise that Brazilian modernists "rediscovered" entire ensembles in the State of Minas Gerais, with Ouro Preto, Mariana, Diamantina, São João del Rey, Tiradentes, and Serro being worthy of mention, among others. In their search for a "deep" national identity, they identified expressions of genuine Brazilian civilization in these eighteenth-century ensembles, and in so doing, they revalued the local Baroque style, which had long been considered eccentric and unimportant. Instead, the modernists claimed the style was a new cultural synthesis, created by an isolated society which had managed to rework various cultural influences in its own manner.

In this sense, the approach of Brazilian modernists to the eighteenth century can be compared with the approach of the European avant-garde to the primitive and archaic – with the particularity that the primitive in Brazil pointed to the country's cultural roots. In Brazil, the avant-garde's rediscovery of primitive culture corresponded to the rediscovery of another, namely a nonofficial national culture, one which had been ignored for years. In terms of architecture, this reading of the national past eventually came to play an important role both in the formulation of Brazil's preservation policy and in Brazil's distinct strain of modernist design.

The combination of a search for the new and a revaluation of tradition is clearly evident in the course taken by Lucio Costa, the creator of Brasília, who was also the intellectual leader of the Brazilian architectural renovation movement in the 1930s. Costa's enduring goal was to integrate modernity and tradition, based on a reflection of the particularity of his professional field, architecture, and its relationship with Brazilian reality (Fonseca, 1997, p.98). In this regard, it should also be pointed out that Brazilian modernist architects had a pragmatic interest in recovering the nation's colonial past, since they believed its traditional forms contained important lessons. Most importantly, they identified a correspondence between colonial and modern architecture, which, in their opinion, shared common characteristics such as simplicity, austerity, purity, and the proper use of materials.

The cultural critic Antoine Compagnon referred to the discourses developed by the modernist avant-garde and within the formalist critique known as "orthodox narratives." According to this view, the course of art within modernity was characterized by a reductive search for essence. While analyzing the critical works of Hugo Friedrich in literature, and Clement Greenberg in the fine arts, Compagnon showed how the development of modern art embodies a "purification dialectics". This was undoubtedly the position of Brazilian modernist architects, who viewed architecture as moving progressively towards an increased authenticity and autonomy, as was reflected in values such as structural truth and simplicity. As such, their reading of tradition represented precisely this an "orthodox narrative". It should then come as no surprise that they rejected the historicist nineteenth century (whose architecture, as well as being "imported," was believed to have been characterized by ornamentation and superficiality), in favor of the austere colonial architecture of the eighteenth century. In their pursuit of depuration, the modernists needed to choose prominent ancestors, and the severe builders of the gold-cycle towns of Minas Gerais corroborated theses defended by Costa, Niemeyer, and others. Thus, as Compagnon pointed out, the "orthodox narratives" of modernism tended to be both teleological (insofar as they reflected the very outcome they intended to arrive at), and apologetic (insofar as they legitimated contemporary production) (Compagnon, 1996, p.44).

Manufacturing Heritage: The Making of a National Symbol

This particularly Brazilian modernist approach to the past raised the important question of the preservation of the nation's colonial heritage, which was seen as essential to the construction of Brazil's national identity. With the "1930 Revolution" the issue also became a main focus of the new ruling group, which tried to establish a government-oriented cultural policy.[5] Towards this end, the new rulers engaged a number of progressive intellectuals from the modernist movement to work in the Ministry of Education and Health. In July 1933, these appointees' first historic-preservation step was to make Ouro Preto a "national monument" by Decree No.22.928. In fact, this action was largely symbolic, since no specific legal provision was taken to protect the site or its individual monuments.

5 The so-called "1930 Revolution" was basically a military revolt that culminated in a coup d'état replacing Washington Luiz with Getúlio Vargas as president. This movement, which inaugurated a new regime, "was made up mostly of young men, both military and civilian, and produced upheaval and fresh beginnings in social and economic life as well as in politics" (Mindlin, 1956, p.4).

But in 1936 the Minister of Education, Gustavo Capanema, with the help of Mário de Andrade, one of the most prominent modern Brazilian writers, prepared a conservation law to be submitted to the Federal Parliament. An even more important step was taken with the creation of a new federal agency to oversee heritage preservation: called *SPHAN – Serviço do Patrimônio Histórico e Artístico Nacional* (National Historic and Artistic Heritage Service) – which was incorporated into the organizational structure of the Ministry of Education.[6] Finally, in 1937, another decree, No.25, provided SPHAN with the legal means for an effective preservation policy, the instrument of *tombamento* (listing). This legal procedure was applied almost immediately to Ouro Preto and other historic centers in Minas Gerais and enabled SPHAN to prevent damage to or demolition of a listed building, in addition to giving it control of introducing new buildings to a listed site.

It is interesting to note that both the mechanism of official government protection and the first measures taken by SPHAN derived directly from the "orthodox narrative" of modernism, as described above. In Brazil this led to the establishment of an affinity between the Baroque architecture of the colonial past and the modern architecture of the present. According to this view, authentic Brazilian architecture had only been developed during the mining cycle of the eighteenth century, and all previous buildings had merely been transplants of Portuguese architecture – a sort of "pre-history" of Brazilian architecture. One corollary of this ideological posture was the myth that only Brazil's Baroque architecture – as well as its contemporary modernist architecture – had dignity, and that the interval of 150 years inbetween had been sterile.

In this way, an "oblivion strategy" was set in motion which stipulated that in order to recall an idealized (austere and authentic) eighteenth century, it was necessary to "forget" the slow evolution that cities such as Ouro Preto had undergone since. Therefore, in the very listing of these historic centers as a preservation site, the artistic value of their

6 It is interesting to note that, although connected to a federal ministry, SPHAN was marked, since its establishment, by a relative autonomy in conducting its affairs. Supported by the work of a capable body of technical employees and a work approach that has remained relatively homogeneous throughout its entire life, it has been characterized by a continuity that is seldom found in public administration in Brazil.

colonial ensembles triumphed over their historical value. However, privileging aesthetic expression in this way eclipsed the towns' actual histories and many of the components that made them a socially constructed whole. Government preservation policies in the years since these cities were first listed have only further emphasized this focus on the colonial ensemble as an idealized object. As an example, Lia Motta has caustically summarized the preservation practice imposed on Ouro Preto as follows: "Economically emptied, the city was used as a raw material for a national laboratory of modernist inspiration, leaving its population, which was not even mentioned, subordinate to that idealized view" (Motta, 1987, p.108).

In his book, *The Past is a Foreign Country*, David Lowenthal asserts that every act of recognition "alters survival from the past" (Lowenthal, 1986, p.263). This seems to be exactly the case with cities like Ouro Preto. In its search for a national symbol, SPHAN initiated an effort to homogenize the image of the city, thereby eliminating the majority of its recent urban and architectural transformations and, concomitantly, important references to local history. In essence, a systematic action was taken to efface traces of the nineteenth century – requiring, for example, the removal of such Neoclassical or eclectic architectural elements, such as gables and platbands as a condition for remodeling approval. Such elements were viewed as disruptive to the colonial ensemble's unified appearance. Examples of such "corrective" actions may be found throughout the city, but one particularly noteworthy example is the *Banco Comércio e Indústria,* where a colonial facade was added, and nineteenth-century architectural elements removed.

Similar homogenizing stylistic criteria were also mandated for new buildings. Initially, these envisioned that some modern buildings would be accepted, provided that they were "of good architectural quality," as evaluated by SPHAN's experts. In setting such a standard, SPHAN was following the directive of Lucio Costa, who (in line with his modernist convictions) argued that sooner or later SPHAN would have to ban "colonial disguises".

Costa also asserted there was nothing worse than the tendency, which he identified as predominant in the United States and Great Britain, to reproduce everything in an "appropriate style," including light switches. Thus, during the initial years of the preservation effort, approval was given, for example, to the well-known Grande Hotel, designed by Oscar Niemeyer. Eventually, this building caused heated arguments among both intellectuals and the local population regarding the evaluative criteria used by the agency. Furthermore, given the difficulty of analyzing such designs, in later years, SPHAN began to employ more prescriptive rules that specified a number of stylistic features, such as typical details of construction and finishing for roofs, cornices and sashes, and strict color schemes. The imposition of such design guidelines for new buildings soon caused a "heritage style" to appear in Ouro Preto, which was typified by contemporary buildings that emulated houses from the eighteenth century. Because the city was not expected to grow rapidly, SPHAN's initial regulatory attention focused on building facades, while disregarding other design parameters such as lot size, house position, and building volume, which would soon prove to be very important (Motta, 1987, p.103). However, when the city did begin to grow rapidly, especially in the 1960s, this approach led to a harmful "counterfeiting" of the ensemble, and the emergence of a hybrid architecture in which "heritage-style" buildings mingled with original specimens. Here, as Lowenthal asserts, "the passage of time dissolves distinction between originals and emulations, and augments their confluence" – which ends up, even from the viewpoint of aesthetic recognition, posing a problem (Lowenthal, 1986, p.160).

In essence, the local history – the intricate web of social, economic and cultural relationships that made up the face of the place and the life of the town – was discarded to provide room for an idealized national symbol. According to such a preservation policy, which systematically excludes participation by local residents, it might come as no surprise if the place memorability of the entire city eventually disappeared – at least as far as the local population is concerned. One might even note how a curious inversion has taken place: in these historic centers the state assumed the role of "local guardian," while local residents came to be viewed as hostile in terms of the preservation agenda, being transgressors within their own city.

Consuming Tradition: Preservation And Development

Considering the above relationships, it is not surprising that state preservation policies have found diminishing support among the local population in the listed historic centers, which tend to consider SPHAN as an exogenous agency whose purpose is to preserve their cities for outsiders. Even the opening of a local SPHAN office in some of the ensembles, such as in Ouro Preto, has not helped to change such perceptions; on the contrary, SPHAN employees are viewed as intrusive and hostile to local interests. It is noteworthy that such antagonism has never grown into open and active opposition, as might be expected.

As late as the 1960s, there was no serious conflict resulting from the divergence of views between SPHAN and the local population. Economically emptied, these cities remained largely unchanged from colonial times. However, this situation changed rapidly, starting in the late 1960s with the region's development of the aluminum and iron mining industry which brought new momentum to the economy of these centers. Just to give an idea of what happened in Minas Gerais at the time, we can present a single figure that gives an idea of the magnitude of this expansion process: when Ouro Preto was listed in 1938, it comprised approximately 1,000 buildings; by contrast, an additional 3,000 structures were approved by 1985.

Such modernization pressures soon resulted in fierce antagonism among the local population, whom had been systematically excluded from the formulation process for preservation policies, and SPHAN, which continued to try to maintain the colonial ensemble's appearance through building-design controls. However, by viewing these historic cities purely as works of art, the preservation agencies instituted policies that cannot reconcile such new pressures. The situation was compounded by the failure of the various government agencies to cooperate on a regional planning effort.

As an example, we can highlight the case of Ouro Preto, where efforts at comprehensive planning were first instituted in the late 1960s, when it became obvious that uncontrolled expansion was causing the city's colonial ensemble to become disfigured. In 1968, Portuguese architect Viana de Lima, a UNESCO consultant, prepared the first such development plan, which primarily consisted of a zoning map and defined areas of preservation and expansion. A few years later, a multidisciplinary team, under the direction of the *Fundação João Pinheiro*, a state-government planning agency, drew up a more comprehensive plan. This effort between 1974 and 1975 included plans for new urban infrastructure and landscaping; the restoration of monuments; and social, economic, institutional and administrative initiatives. In addition, an urban expansion scheme was prepared, proposing the creation of new centers to ensure that development could continue in a consistent manner, without affecting the integrity of the colonial ensemble (Castriota, 2009).

It is indicative of the problems of Ouro Preto that almost 25 years after such plans were first proposed, institutional difficulties continue to block their implementation. Much of this delay may be attributed to a lack of cooperation between the various agencies responsible for the preservation and administration of Brazilian cities. For example, despite federal, state and local authorities being involved in Ouro Preto, they do not always act in a collaborative manner. With respect to heritage preservation, Brazilian legislation provides for "concurrent competence" of the federal union, states, and municipalities, but it fails to define how each level of government is to act. Thus, in Ouro Preto, SPHAN (a federal agency) is in charge of listing procedures, and has been responsible for the maintenance and conservation of the city since 1938 – including the inspection and coordination of designs and maintenance of museums. The state government, represented by the *Fundação João Pinheiro,* has been charged with overall planning for the city and its outlying region. The city council maintains its role as the principal local administrative power, responsible for, among other things, enforcing the zoning laws. Conflicts have become quite common in this atmosphere of competing spheres of power and interest. For example, as a representative of local interests, the city government has often opposed SPHAN policies, which it regards as unnecessarily restrictive. In the opinion of an architect from SPHAN, "the local authority used to omit their responsibility for the conservation of local heritage and even make the work of SPHAN more difficult". The local government has also failed to support the urban planning efforts of the state government, resulting in these efforts having generally failed to have any impact on urban conditions. As for SPHAN, despite the incorporation of new concepts into its discourse (especially the "Declaration of Venice," advocating such new ideas as the socially oriented use of monuments), it has continued to consider the historic centers as primarily aesthetic, as objects to be preserved, and continued to ignore matters of economic growth and social development.

References

AMARAL, A. Stages in the Formation of Brazil's Cultural Profile. *The Journal of Decorative and Propaganda Arts*, 21. Miami: DAPA, 1995.

ARANTES, O. *Chai-Na*. Editora da Universidade de São Paulo, São Paulo, 2011.

BOMENY, H. *Guardiães da razão*: modernistas mineiros. Rio de Janeiro: Editora UFRJ/Tempo Brasileiro, 1994.

CASTRIOTA, L. *Patrimônio cultural*: conceitos, políticas, instrumentos. São Paulo, Annablume, 2009.

COMPAGNON, A. *Os cinco paradoxos da modernidade.* Belo Horizonte: Editora da UFMG, 1996.

FONSECA, M. C. L. *O patrimônio em processo*: trajetória da política federal de preservação no Brasil. Rio de Janeiro: Editora UFRJ/Minc-IPHAN, 1997.

LOWENTHAL, D. *The Past is a Foreign Country.* Cambridge: Cambridge University Press, 1986.

MINDLIN, H. E. *Modern Architecture in Brazil.* Rio de Janeiro/ Amsterdam: Colibris Editora Ltd., 1956.

MOTTA, L. A SPHAN em Ouro Preto: uma história de conceitos e critérios. *Revista do Patrimônio Histórico e Artístico Nacional.* Brasília: Secretaria do Patrimônio Histórico e Artístico Nacional: Fundação Pró-Memória, 1987.

STOVEL, H. Approaches to managing urban transformation for historic cities. In: LUNG, D. (Ed.). *The Conservation of Urban Heritage*: Macao Vision, Macao, 2002. p.103-20.

WANG, G. *Voices in Cultural Heritage Preservation*: a comparative study between New Orleans and Changting (China). Graduate Faculty of the Louisiana State University, 2007.

WONG, R. W. M. *Preservation of Traditional Buildings with Heritage Value in Asian Cities with Colonial Background* – Shanghai, Guangzhou and Hong Kong cases. (Available at: http://personal.cityu.edu.hk/~bswmwong/pp/ckc&fwfront.html.)

26

Risk and Climate Change in Brazilian Coastal Cities (The case of São Paulo State)

Leila da Costa Ferreira[1]

Introduction

Over a short period of time, climate change has become one of the most pressing issues in the 21st century (Giddens, 2009; Hogan, 2009a). Since 1988, the Intergovernmental Panel on Climate Change (IPCC), established by the United Nations (UN) and by the World Meteorological Organization (WMO), has been engaged in understanding the functioning of the global climate system. The greatest scientific advances of this movement came with the most recent reports, when there were relative consensus expressed regarding global warming by concluding with high confidence that climate change is unequivocal, and that the increase in the average temperature of the Earth's surface results from human activities combined with natural variation of the global climate system (IPCC, 2007). This fact represents an advance in the debate since previous positions and arguments persisted by defending, with some power of public persuasion, that contemporary

1 Full Professor, Institute of Philosophy and Human Sciences (IFCH), Centre of Environmental Studies and Research (NEPAM), University of Campinas (UNICAMP). Brazil. E-mail: *leilacf@unicamp.br*.

global warming was a result of natural causes from biogeological planet cycles. According to these positions, with it being a natural problem, there would be nothing to do about it apart from accepting and coping with it (Giddens, 2009).

Although the scientific community recognises that the Earth's temperature is subject to long-term cyclical variation, what is being argued is that it is possible to assess the contribution of human activities to the so-called "expanded greenhouse effect" and observe that the pace of this phenomenon has been verified since the advent and internationalisation of the industrial revolution and western life style, in more or less accelerated rhythms in several countries (Wilbanks; Kates, 1999; Beck, 2000; Pittock, 2005). The changes currently being experienced, apparently unprecedented, challenge the adaptive and responsive capacity of the planet since they are still largely unknown (Young et al., 2006), despite several scientific initiatives having sought to provide a better understanding of these processes on an integrated planetary scale (i.e. Rockström et al., 2009).

Thus, living in a global society means facing multiple risks shaped by multiple stressors, as there are great uncertainties in terms of what the world will be like in 20, 30 or 40 years. It is only possible to mention probabilities and possible scenarios. Based on this, the highly modernized society is involved with the challenge of managing increasing risks. In terms of responses to climate change, it has been argued elsewhere that governments are important actors that play an important role in setting regulations, institutions and appropriate modes of governance for confronting these risks at different levels and scales (Giddens, 2005, 2009; Bulkeley; Kern, 2006; Alber; Kern, 2008; Newell; Bulkeley, 2010).

In Brazil, where major cities are located in coastal zones, the impacts of climate change are an extremely relevant issue. Coastal areas are considered particularly vulnerable to climate change (Adger, 1999; Nicholls et al., 2007); at the same time, they concentrate an expressive contingent of the world's population (McGranahan et al., 2007), and are development centers of the worldwide economy due to their infrastructure and natural resources (Satterthwaite et al., 2007). In this sense, this chapter reflects on the challenges of climate change for these areas, how the issue could be framed and how local governments in Brazilian coastal cities in the State of São Paulo are prepared (or not) to deal with this risk in terms of policy strategies and instruments (Figure 1).

Figure 1. Coastal cities in the State of São Paulo, Brazil.
Source: Elaborated by Allan Iwama 2010.

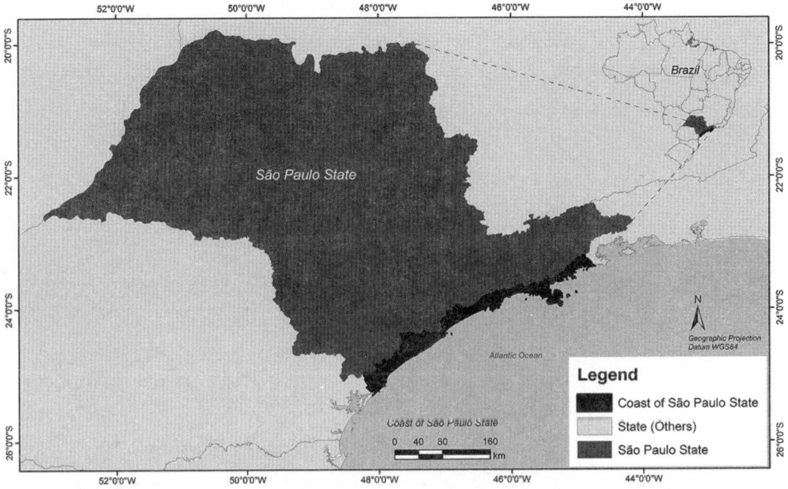

Despite this debate being in its early stages in Brazil, this chapter represents the first attempt to investigate some of the political-institutional structures that can have some kind of effect on relevant areas for intervention in terms of climate change at a local level in the 16 coastal cities of the State of São Paulo.

Risk and Climate Change

A reconsideration of the nature of the current ecological crisis is suggested by Beck (1992, 1995, 2009), since the side effects of industrial production are characterized as a deep institutional crisis of industrial society itself. This proposition can be applied to climate change, whose causes are deeply rooted in modern societies: apparently innocuous and invisible gases such as methane (CH_4) and carbon dioxide (CO_2) are released into the atmosphere as development products (industrial agriculture, increased energy consumption and economic growth), thereby altering its composition with incalculable consequences (Bulkeley, 2001).

In the risk society, the technical-industrial development causes unpredictable, invisible, incalculable risks that require society's self-reflection, self-confrontation in relation to the basis of social cohesion, and an examination of its rational foundations. The risk that characterises modern industrial society is closely related to the future, to the misfortunes actively assessed in relation to future possibilities and situations with which our society has had little experience confronting (Beck, 1992, 2009; Giddens, 2005, 2009). These characteristics are closely related to climate change risks and are influenced by the increasing process of globalization (Young et al., 2006).

It is hard to understand the level of these risks and, in many cases, it is impossible to predict them before it is too late, since we do not know what other changes will come, or what dangers they will bring with them. Thus, the challenge posed is how to establish effective policies to address different problems, since the future is uncertain and it is not possible to evaluate the risks with complete accuracy until the anticipated hazards eventually materialise (Giddens, 2009). The required political action and intervention at the international, national and local levels will have a decisive effect in setting the boundaries to avoid dangerous global warming and support adaptation efforts that are an increasing occurance (Parry et al., 2008; Parry, 2009), particularly in the developed countries (i.e. Moser 2009 for USA; Hulme and Neufeldt 2010 for Europe).

Climate Change and Local Governments

Global changes in climate, environment, economy, societies, governments, institutions and cultures converge in certain locations. The effects at the local level, in turn, contribute to global changes and are affected by them. As a result, connecting the local and global scales across a broad range of disciplines and problems – integrated assessments of population, economy, technology and environmental changes – enables a potentially deeper understanding of the global environmental change, including climate change, in all its complexity (Wilbanks; Kates 1999; Betsill; Bulkeley, 2007; Okereke et al., 2009).

According to data from several countries, local governments are able to control policies that can deal with about 30-50% of national greenhouse gas (GHG) emissions (Robinson; Gore, 2005). Local governments are usually responsible for local transportation, development planning and energy management (Coenen; Menkveld, 2002; Bulkeley; Betsill, 2003). Since a large number of people live and work within city boundaries and are under local government jurisdiction, these governments are in a privileged position to influence many activities that can generate large GHG emissions (Yienger et al., 2002).

At the same time, with urbanisation, a centralised state is less necessary, since the urban population can solve most of their problems locally. This reality has led the most developed countries to adopt a state structure with broad participation from local governments. In addition, a part of the literature that analyses sustainability frameworks (Castells, 1983; Cahn, 1995; Buttel, 2000) clearly indicates the importance of decentralization, that is, the participation of several social actors and stakeholders in finding solutions to environmental problems, and, at the local level, where opportunities are more noticeable.

In this sense, the quality of local government has an important role in responding to climate change risks (Satterthwaite et al., 2007). Currently, many local governments in developing countries have a limited institutional and response capacity in dealing with the several problems that affect the quality of life of these populations, especially in terms of the poorest that have less coping capacity and weak ties in social protection networks (Satterthwaite et al., 2007; Tanner et al., 2008). However, climate change does not only reach the poor, its impacts affect different sectors of society, the public sphere and also the market (Beck, 1992, 2009; Wilbanks et al., 2007; Satterthwaite et al., 2007; Tanner et al., 2008; Satterthwaite, 2008; Bartlett et al., 2009). Local governments are considered critical elements to address many of the causes and consequences that arise from climate change as they are close to wherein these activities and events take place (Wilbanks; Kates, 1999; Robinson; Gore, 2005; Satterthwaite et al., 2007; Puppim de Oliveira, 2009).

A summary of some local and state governments' responsibilities in areas related to climate change and risk management, which must be components of a comprehensive adaptation strategy to climate change, is presented in Table 1. Usually local governments are responsible for:

- Finance: financial and accounting management of municipal budget; taxes collection and management, licenses and fees.
- Engineering and public construction: construction and maintenance of the public space.
- Urban/local development: land use regulation, urban zoning, real state registration and urban planning.
- Health and public hygiene: collection, distribution and treatment of drinking water, pollution control, collection and treatment of solid waste, sanitary hygiene, public areas cleaning, medical and ambulatory services.
- Social urban policies: housing, schools, day-care centres, youth, elderly, etc.
- Civil defence and emergency response: disasters relief, fire brigade, ambulance services and rescue.
- Public administration and human resources: various administrative tasks and responsibilities including human resources management.

Table 1. Role of Local Governments in Areas with Effects on Climate Change

Local government role	Long term planning	Prevention of disasters	Response to extreme events	Reconstruction
Built environment				
Construction Code	High	n.d.	High	High
Land use regulation and real estate registration	High	Low	n.d.	High
Maintenance and construction of building and public space	High	Low	n.d.	High
Urban Planning and Zoning	High	n.d.	High	High
Infrastructure				
Water treatment	High	Low	High	High
Wastewater treatment	High	Low	High	High
Drainage system	High	Low	High	High
Streets, avenues, bridges	High	n.d.	High	High
Electricity	High	Low	High	High

Local government role	Long term planning	Prevention of disasters	Response to extreme events	Recons-truction
Landfill	High	Low	n.d.	High
Services				
Protection against fire	High	Low	High	Low
Public order, police	Average	High	High	Low
Solid waste collection	High	High	High	High
Education	Average	Average	n.d.	n.d.
Health and public hygiene	Average	Average	High	High
Public Transportation	Average	High	High	High
Social welfare	Average	High	High	High
Responses to disasters and civil defense	n.d.	n.d.	High	High

Source: Adapted from Satterthwaite (2008, p.26).

Climate Change in Brazilian Coastal Cities

The Coast of the São Paulo State

In Brazil, many important cities and infrastructure are located near the coastline, which makes climate change an extremely relevant issue for the present and future of these areas (Nicolodi; Petermann, 2010). It is also estimated that more than 20% of the Brazilian population live in coastal cities (Zamboni; Nicolodi, 2008). In this context, São Paulo – the largest Brazilian state in terms of population concentration and economic development – has much of its infrastructure located on its coastal region. The ports of Santos and São Sebastião, respectively the biggest port in South America and the main distribution centre for Brazilian inland oil, along with the industrial district of Cubatão and the tourism industry, are just some examples of the importance of this region regarding the social and economic development of Brazil.

The coast of São Paulo covers about 700 km comprising 16 cities with over two million inhabitants (see Table 2). There are diversified ecosystems that are home to estuaries, mangroves, sandbanks, caves, rivers, rocky shores and about 150 islands (SMA 2005). Moreover, it has several protected areas within its territory and contains the most important remnants of the Atlantic Forest, one the three most threatened ecosystems on the planet and a hotspot for biodiversity (Aidar et al., 2001). Although there are a number of measures in place to protect these important ecosystems, they have been endangered by increasing tourism activities, modern transportation infrastructures, oil and gas industry, and illegal exploitation of natural resources (Borelli, 2006).

In terms of regional and socio-economic aspects, the Northern section of the coast of the State of São Paulo is currently being targeted by a series of development interventions and expressive investment in infrastructure, especially in the area that corresponds to the cities of Caraguatatuba and São Sebastião (Hogan, 2009b). In this sense, the construction of a number of new facilities is expected in the coming years:

- a road bypass connecting the cities of São Sebastião and Caraguatatuba;
- a regional landfill to minimise the chaotic situation of solid waste in the region;
- a prison facility in Caraguatatuba;
- a natural gas treatment plant (already in an advanced stage of construction).

Despite the absence of consolidated data, it is expected that this region will be affected by climate change, as well as by the ongoing interventions and projects, whose results might lead to a second phase of disruptive urban growth and expansion, with consequences that could be drastic for the integrity of the Atlantic Forest and for the different social groups, especially the local populations that live in the area (Borelli, 2006; Marengo, 2007; SMA, 2009; Hogan, 2009b). In this context, it is appropriate to consider how climate change can affect this area and the links between these processes. Far from finding exhausting answers, we take the mapping of public agencies, municipal bodies and specific institutions as a starting point, whose main focus is to act on environmental issues in order to assess the existence of an institutional capacity that could be mobilised in the future to address climate change in these areas.

Political-institutional strategies

Regional level strategies

In Brazil, the city of São Paulo was a pioneer in approving a specific law to deal with climate change. The municipal law (n. 14933) was approved in 2009 and establishes the reduction of 30% in total GHG emissions of the city. It enforces public transportation priority, gradual reduction in fossil fuel consumption and dependence, a compulsory recycling program and reduction of solid waste together with energy efficiency criteria combined with eco-building.

Some months after this law's approval, the State of São Paulo also approved its own law becoming the second Brazilian state to approve such type of regulation (the first one being the State of Amazonas in 2007), which also presents tangible mitigation targets. The so-called State Climate Change Policy expects a reduction of at least 20% in total CO_2 emissions by 2020 compared with the 2005 level. Every five years, until 2020, the government will be able to set intermediate targets to reach this goal of 20%.

In spite of being considered substantial advances to address the climate change problem, these laws rely on a network of political-institutional structures and governance arrangements in order to be effectively applied (see Table 1). Thus, it is fundamental to investigate the presence of some of these structures in the 16 coastal cities of São Paulo State as an inherent part of our research agenda, as these cities do not have any clear climate change response in terms of specific laws or policies as in the case of the city and the state of São Paulo. Hence, in order to assess the institutional and administrative capacity of these coastal cities to internalise the climate change issue, a mapping exercise was carried out aiming to find the presence of governmental bodies and institutions interlinked with sectors and policies with some level of effect and jurisdiction to address climate change.

Local political-institutional structure

Our mapping exercise focused on the executive and the legislative municipal bodies as well as on the presence of secretariats, departments, other public agencies, and City Council commissions that have some history and capability in dealing with areas that can affect or be affected by climate change. As these coastal cities do not have particular institutions to respond to climate change so far, the idea of looking at the municipal realities was used as a strategy to frame the climate change issue within the current administrative structures. Despite recognizing that this exercise is not exhaustive and that the presence of these structures and institutions that work with different sectors does not guarantee neither the effectiveness nor the adequate response to climate change, it is intended to be a starting point for understanding the capacity of these coastal cities in terms of their ability to formulate and implement climate strategies.

Table 2 shows that a large number of the investigated cities have political-institutional structures that could be mobilised to address the climate change issue, especially in terms of specific secretariats, governmental agencies and departments. Our survey highlights the presence of structures dedicated to social, economic and urban development policies, besides those regarding the environment and health.

Table 2. Political-Institutional Structure in Coastal Cities of the State of São Paulo

Area of action*	List of Cities	Number of Cities
Economic Development: Secretariats; Authorities (Agriculture; Fishery; Industry; Commerce; Science; Technology)	Cananéia, Caraguatatuba, Cubatão, Guarujá, Ilha Comprida, Itanhaém, Mongaguá, Peruíbe, Santos, São Sebastião, São Vicente, Ubatuba.	12
Social Development: Secretariats; Authorities (Social Action; Social Promotion; Social Assistance; Citizenship)	Bertioga, Cananéia, Caraguatatuba, Cubatão, Guarujá, Ilhabela, Ilha Comprida, Itanhaém, Mongaguá, Peruíbe, Praia Grande, Santos, São Vicente, Ubatuba.	14
Urban Development: Secretariats; Authorities (Housing; Public Construction; Public Services; Infrastructure)	Bertioga, Cananéia, Caraguatatuba, Cubatão, Guarujá, Iguape, Ilhabela, Ilha Comprida, Itanhaém, Mongaguá, Peruíbe, Praia Grande, Santos, São Vicente, São Sebastião, Ubatuba.	16

Area of action*	List of Cities	Number of Cities
Environment: Secretariats; Authorities	Bertioga, Cananéia, Caraguatatuba, Cubatão, Guarujá, Iguape, Ilha Bela, Itanhaém, Mongaguá, Peruíbe, Praia Grande, Santos, São Sebastião, São Vicente, Ubatuba.	15
Planning (Management): Secretariats; Authorities	Caraguatatuba, Cubatão, Guarujá, Ilha Comprida, Itanhaém, Mongaguá, Peruíbe, Santos, São Vicente, Ubatuba.	10
Health: Secretariats; Authorities	Bertioga, Cananéia, Caraguatatuba, Cubatão, Guarujá, Iguape, Ilha Bela, Ilha Comprida, Itanhaém, Mongaguá, Peruíbe, Praia Grande, Santos, São Sebastião, São Vicente, Ubatuba.	16
Security: Secretariats; Authorities; Directories (Public Security; Urban Security; Social Defense; Civil Defense)	Caraguatatuba, Guarujá, Itanhaém, Mongaguá, Peruíbe, Santos, São Sebastião, São Vicente, Ubatuba.	9
Transport: Secretariats; Authorities	Caraguatatuba, Iguape, Itanhaém, Mongaguá, São Vicente.	5

* Some secretariats, authorities or directories have a specific area of intervention (for example Planning; Environment, etc.), while others act in more than one area at the same time (Planning and Environment, for example).

Source: The authors.

Regarding the legislative body, most of the analyzed City Councils have commissions whose interest will be able to intervene and govern by authority, basically through legislation, such as through law ones approved in the city and in the State of São Paulo, that aims to mitigate GHG emissions, as well as create and mobilize the necessary conditions and resources for adaptation strategies.

The area of Planning and Management does not possess a specific structure in some of the analyzed cities, which might cause problems for coordinating inter-sectoral and multidimensional actions. In the case of Transport, only five cities directly manage the public transport in their localities. In the other cities, transport is indirectly administrated through autarchies and public companies. In many cities, the transport sector is one of the main sources of greenhouse gases emission; thus, having ways to govern this sector is fundamental for mitigation actions (Collier, 1997; Robinson; Gore 2005; Bulkeley et al., 2009). In terms of environmental policies at the local level, only the city of Ilha Comprida does not possess a specific structure to deal with this area, although these activities can be incorporated within its urban development department.

These data may suggest that the studied local governments present the first steps for internalizing the climate change issue in their government agendas. Despite the presence of such structures and institutions being considered important, the implementation of actions up to now are far from those required to face the expected challenges from the scenarios projected by the scientific community (IPCC 2007; Parry et al., 2008; Parry, 2009).

Local strategies on climate change

The implementation of local policies regarding climate change requires the integration of several areas (Table 1) and may stand against many challenges posed by traditional political practices and institutions. Some pre-existing patterns of political intervention are sometimes responsible for the inability to assess and formulate complex interactions among the sectors involved in climate change policies (Murgida; Natenzon, 2009).

In the coastal cities of the State of São Paulo, local authorities have argued that many problems to deal with such issues arise from a lack of resources as well as advanced political instruments and alternative interactions among offices and departments. The sectoral and social interests at stake and the existing political capacities provide a narrow space, over the long-term, and far reaching climate policy implementation.

Even though science is advancing to provide increasingly reliable data, methodologies, tools and frameworks to help society cope and adapt to the impacts of climate change and variability, these local governments located on the coast of the State of São Paulo are still struggling to provide the most basic urban services for a proportion of their inhabitants in terms of adequate housing, sanitation and health care. The lack of financial and human resources, combined with the high level of uncertainty in regards to climate change impacts (Beck, 1992, 2009; Giddens, 2009), and the need for climate change scientific data at local level, also contribute to the lack of consistent action towards climate change risks in this region.

The communication between specific offices and departments with civil society organizations is another aspect that deserves attention. Some of these cities have strong social movements regarding environmental issues, including in terms of fishing, forest preservation and others. In some cities, the participation of such groups is quite expressive in confronting many damaging initiatives. Despite this, there are no coordinated policies that connect these groups with local authorities; thus, these liaisons remain occasional and voluntary opportunities. The local Management Councils, which are deliberative public spaces, are not very active in these cities but represent an important space where policies can be designed and evaluated with the participation of several sectors of the society.

Even though these cities do not have climate specific strategies, the fact that a general environmental agenda is institutionalized through some type of legislation in most of the cities (Table 4) may suggest possible ways to internalize climate change issues within local governance. In this sense, the political-institutional structure is favourable; however, it is necessary to think of communication strategies among existing policies, as well as local and state-level measures in areas to some extent related to climate change.

Final Comments

According to the risk theory discussion covered in Beck (1992, 2009), the risks brought by climate change are characterized by escaping the perception, situated in the sphere of physical and chemical formulas, of remaining invisible. They can be changed, magnified, dramatized or minimized within knowledge and, accordingly, are open to social definition and construction. Also, these risks are closely related to the future and the misfortunes that are actively assessed in relation to future possibilities.

In this scenario, political action and intervention at international, national and local levels will have a decisive effect on establishing limits to global warming and on adapting to what is already in progress (Giddens, 2009). International agreements are essential, and other agencies, including NGOs and the private sector also have key roles; however, the state has many powers that have to be raised, if more serious impacts involving global warming occur, as previously discussed in section 3.

In terms of climate change policies, the debate in Brazil is still in its early stages and there are a number of issues to be worked through in the coming years. Despite being a new subject in the political arena, legislation exists that has been approved since 2007 at the national, state and municipal level that aim to mitigate GHG emissions and coordinate adaptation efforts, which are combined with risk management and disaster risk reduction, as previously presented in section 4.

The fact that local governments are considered critical actors to address many of the causes and consequences that arise from climate change was also discussed, as they are close to wherein these activities and events take place, considering areas such as built environment, infrastructure and services, as shown in Table 1, which have direct and indirect effects on climate change.

This chapter reflects the first attempt to investigate some of the political-institutional structures that have some effect on relevant areas for intervention, in terms of climate change at the local level, in the 16 coastal cities in the State of São Paulo. Such analysis (Tables 2 and 3) shows that these cities have secretariats, departments, directories and other types of governmental bodies, as well as environmental legislation that will have to be mobilized when undertaking climate change actions.

We have argued that the existing structures, and the alternatives that have being proposed up until now by these institutions, are far from sufficient to respond effectively to the magnitude and complexity of this problem. If we consider that we are debating and deciding on the future – even though the future has been made present – and the quality of the life in the planet, not only for humans but also for diverse ecosystems, depend on actions and measures taken now, then the socio-environmental (in)justice that devastates millions raises ethical dilemmas. Emerging from this debate are reasonable argu-

Table 3. Environmental legislation in the Coastal Cities of the State of São Paulo

Cities	Municipal Organic Law	Land Use Planning	Environment Code	Urban Development Planning	Regional Ecological-Economic Zoning	Conservation Units Acts	Others
Bertioga		X	X				
Cananéia							X
Cubatão	X						
Guarujá	X	X					
Ilha Comprida	X						
Itanhaém	X	X		X			
Peruíbe	X						
Praia Grande		X		X	X		
Santos	X	X				X	X
São Vicente		X					X
São Sebastião							X
Ubatuba	X						

Source: The authors.

ments to defend the engagement in collective and political dialogues, negotiation and proposals that could enhance the response capacity to the challenges that are posed, so that these responses would not be palliative responses that go towards the social and economic maintenance of the status quo, the political interests and the unsustainable patterns of human development.

Thus, we look for the first step and a contribution to the Brazilian scientific community towards understanding the capacity that these specific cities will have to deal with the climate change impacts building upon some existing political-institutional structures. However, some questions still remain unanswered and will have to be address by in future research. How do local and regional actors react to climate change issues? To what extent do local political actors have a complete view of the changes that will happen or are already happening? What makes some cities better prepared to respond to this challenge than others?

References

ADGER, W. N. Social Vulnerability to Climate Change and Extremes in Coastal Vietnam. *World Development*, 27, 2, p.249-69, 1999.

AIDAR, M. P. M. et al. Atlantic Forest sucession over calcareous soil, Parque Estadual Turístico do Alto Ribeira – PETAR, SP. *Revista Brasileira de Botânica*, 24, 4, p.455-69, 2001.

ALBER, G.; KERN, K. Governing Climate Change in Cities: Modes of Urban Climate Governance in Multi-level Systems. *OECD International Conference*, Competitive Cities and Climate Change, 2nd Annual Meeting of the OECD Roundtable Strategy for Urban Development, 9-10 October, Milan, 2008.

BARTLETT, S. et al. Social Aspects of Climate Change in Urban Areas in Low- and Middle Income Countries. *World Bank Urban Research Symposium*, Marseille, 2009.

BECK, U. *World at risk.* Cambridge: Polity Press, 2009.

______. *What is globalization?* Cambridge: Polity Press, 2000.

BECK, U. *Ecological Politics in an Age of Risk.* Cambridge: Cambridge University Press, 1995.

______. *Risk Society.* Beverly Hills, SAGE, 1992.

BETSILL, M.; BULKELEY, H. Looking Back and Thinking Ahead: A Decade of Cities and Climate Change Research. *Local Governments*, 12, 5, p.447-56, 2007.

BORELLI, E. *Cidade e natureza*: análise da gestão ambiental da zona costeira do litoral norte paulista. São Paulo, 2006. Tese (Doutorado em Ciências Sociais) – PUC/SP.

BULKELEY, H. Governing climate change: the politics of risk society? *Transactions of the Institute of British Geographers*, 26 (4), p.430-47, 2001.

BULKELEY, H.; BETSILL, M. M. *Cities and Climate Change*: Urban Sustainability and Global Environmental Governance. London: Routledge, 2003.

BULKELEY, H.; KERN, K. Local Government and the Governing of Climate Change in Germany and UK. *Urban Studies*, 43, 12, p.2237-59, 2006.

BULKELEY, H. et al. Cities and Climate Change: The role of institutions, governance and urban planning. *World Bank Urban Research Symposium*, Marseille, 2009.

BUTTEL, F. H. Sociologia ambiental, qualidade ambiental e qualidade de vida: algumas observações teóricas. In: HERCULANO, S.; PORTO, M. F. S.; FREITAS, C. M. (Eds.) *Qualidade de vida & riscos ambientais*. Niterói: EDUFF, 2000.

CAHN, M. A. *Environmental Deceptions*. The Tension Between Liberalism and Environmental Policymaking in the United States. New York: State University of New York, 1995.

CASTELLS, M. *The city and the Grassroots*. A cross-cultural Theory of Urban Social Movements. Berkeley/Los Angeles: University of California Press, 1983.

COENEN, F. M.; MENKVELD, M. The role of local authorities in a transition towards a climate-neutral society. In: KOK, M. T. J. et al. (Eds.). *Global Warming & Social Innovation. The Challenge of a Climate-Neutral Society.* London: Earthscan, 2002.

COLLIER, U. Local authorities and climate protection in the European Union: Putting subsidiary into practice? *Local Environment*, 2, 1, p.39-57, 1997.

GIDDENS, A. *The Politics of Climate Change*. Cambridge: Polity Press, 2009.

GIDDENS, A. *Mundo em descontrole*. O que a globalização está fazendo de nós. Rio de Janeiro: Record, 2005.

HOGAN, D. J. População e mudanças ambientais globais. In: HOGAN, D. J.; MARANDOLA JR., E. (Eds.) *População e mudança climática*: dimensões humanas das mudanças ambientais globais. Campinas: Ed. Unicamp, 2009a.

HOGAN, D. J. (Coord.) Urban Growth, Vulnerability and Adaptation: social and ecological dimensions of climate change on the Coast of São Paulo. *Research Project approved by the Fapesp Global Climate Change Programme*, 2009b.

HULME, M.; NEUFELDT, H. *Making Climate Change Work for Us*: European Perspectives on Adaptation and Mitigation Strategies. Cambridge: Cambridge University Press, 2010.

IBGE (Instituto Brasileiro de Geografia e Estatística). Brasil. *População*. 2009. Available at: http://www.ibge.gov.br/paisesat/main.php. Accessed in: 2 October, 2010.

IPCC (INTERGOVERNMENTAL PANEL ON CLIMATE CHANGE). *Climate Change 2007: Impacts, Adaptation and Vulnerability*. Contribution of Working Group II to the Fourth Assessment Report of the International Panel on Climate Change. Cambridge: Cambridge University Press, 2007.

MARENGO, J. A. *Caracterização do clima no século XX e cenários no Brasil e na América do Sul do século XXI derivados dos modelos de clima do IPCC*. Brasília: Ministério do Meio Ambiente, 2007.

MCGRANAHAN, G., BALK, D.; ANDERSON, B. The rising tide: assessing the risks of climate change and human settlements in low elevation coastal zones. *Environment & Urbanization*, 19, 1, p.17-37, 2007.

MOSER, S. C. *Good Morning America!* The Explosive Awakening of the US to Adaptation. Charleston and Sacramento, NOAA and California Energy Commission, 2009.

MURGIDA, A. M.; NATENZON, C. E. Social downscalling: a few reflections on adaptation in urban environments. In: DIAS, P. L.

et al. (Eds.). *Public Policy mitigation and adaptation to climate change in South America.* São Paulo: IEA/USP, 2009.

NEWELL, P.; BULKELEY, H. *Governing Climate Change*. New York: Routledge, 2010.

NICHOLLS, R. J. et al. Coastal systems and low-lying areas. In: PARRY, M. L. et al. (Eds). *Climate Change 2007*: Impacts, Adaptation and Vulnerability. Contribution of Working Group II to the Fourth Assessment Report of the Intergovernmental Panel on Climate Change. Cambridge: Cambridge University Press, 2007.

NICOLODI, J. L.; PETERMANN, R. M. Mudanças Climáticas e a Vulnerabilidade da Zona Costeira do Brasil: Aspectos ambientais, sociais e tecnológicos. *Revista da Gestão Costeira Integrada*, 10, 2, p.151-77, 2010.

OKEREKE, C.; BULKELEY, H.; SCHROEDER, H. Conceptualizing Climate Governance Beyond the International Regime. *Global Environmental Politics*, 9, 1, p.58-78, 2009.

PARRY, M. L. Climate change is a development issue, and only sustainable development can confront the challenge. *Climate and Development*, v.1, n.1, p.5-9, 2009.

PARRY, M. L. et al. Squaring up to reality. *Nature Reports Climate Change*, v.2, p.1-3, 2008.

PITTOCK, A. B. *Climate Change*: turning up the heat. London: Earthscan, 2005.

PUPPIM DE OLIVEIRA, J. A. The implementation of climate change related policies at the subnational level: an analysis of three countries. *Habitat International*, 33, p.253-59, 2009.

ROBINSON, P. J.; GORE, C. D. Barriers to Canadian Municipal Response to Climate Change. *Canadian Journal of Urban Research*, 14, 1, p.102-20, 2005.

ROCKSTRÖM, J. et al. A safe operating space for humanity. *Nature*, v.461, 2009.

SATTERTHWAITE, D. *Climate Change and Urbanization*: Effects and Implications for Urban Governance. United Nations Expert Group Meeting on Population Distribution, Urbanization, Internal Migration and Development, UN/POP/EGM-URB/2008/16, New York, 2008.

SATTERTHWAITE, D. et al. Adapting to Climate Change in Urban Areas: The possibilities and constraints in low- and middle-income nations. Discussion Paper N.1. *International Institute for Environment and Development* (IIED), London, 2007.

SMA (SECRETARIA DE MEIO AMBIENTE). *Planejamento Ambiental. Zoneamento Ecológico-Econômico – Litoral Norte*. São Paulo: SMA/CPLEA, 2005.

SMA (SECRETARIA DE MEIO AMBIENTE). *Litoral Norte*. Metodologia para Avaliação Ambiental Integrada de Projetos. São Paulo: SMA/CPLA, 2009.

TANNER, T. M. et al. Urban Governance for Adaptation: Assessing Climate Change Resilience in Ten Asian Cities. *IDS Working Paper*, 315, Institute for Development Studies (IDS), Brighton, 2008.

WILBANKS, T.; KATES, R. W. Global Change in Local Places: How Scales Matters. *Climatic Change*, 43, p.601-28, 1999.

WILBANKS, T. et al. Industry, Settlements and Society. In: PARRY, M. L. et al. (Eds.). *Climate Change 2007*: Impacts, Adaptation and Vulnerability. Contribution of Working Group II to the Fourth Assessment Report of the Intergovernmental Panel on Climate Change. Cambridge: Cambridge University Press, 2007.

ZAMBONI, A.; NICOLODI, J. L. *Macrodiagnóstico da Zona Costeira e Marinha do Brasil*. Brasília: Ministério do Meio Ambiente (MMA), Secretaria de Mudanças Climáticas e Qualidade Ambiental, 2008.

YIENGER, J.; BROWN, L.; SKINNER, N. *Experiences of ICLEI's Cities for Climate Protection Campaign (CCP)*: a focus on Asia. Proceedings of IGES/APN Mega-City Project. Institute for Global Environmental Strategies, 2002.

YOUNG, O. et al. The globalization of socio-ecological systems: an agenda for scientific research. *Global Environmental Change*, 16, p.304-16, 2006.

SOBRE O LIVRO

Formato: 14 x 21 cm
Mancha: 25 x 45 paicas
Tipologia: Times New Roman 11/13
Papel: Offset 75 g/m² (miolo)
Cartão Supremo 250 g/m² (capa)

1ª edição: 2016

Equipe de realização

Edição de Texto
Tikinet (Preparação de original e revisão)

Editoração Eletrônica
Vicente Pimenta

Capa
Megaarte Design

Impresso por :

Graphium
gráfica e editora

Tel.:11 2769-9056